Faithfully yours
Uncle Henry

Uncle Henry Wallace:

Letters to Farm Families

Edited by Zachary Michael Jack

Purdue University Press
West Lafayette, Indiana

Library of Congress Cataloging-in-Publication Data

Wallace, Henry, 1836-1916.
 Uncle Henry Wallace : letters to farm families / edited by Zachary Michael Jack.
    p. cm.
  ISBN 978-1-55753-493-4
  1. Wallace, Henry, 1836-1916--Correspondence. 2. Farm life--Iowa--History--Sources. 3. Rural families--Iowa--History--Sources. 4. Boys--Iowa--Conduct of life--History--Sources. 5. Iowa--Social life and customs--Sources. 6. Agricultural journalists--Iowa--Correspondence. 7. Farmers--Iowa--Correspondence. 8. Teachers--Iowa--Correspondence. 9. Iowa--Biography. I. Jack, Zachary Michael, 1973- II. Title.

  F621.W183A3 2008
  977.7'031--dc22
                            2008003490

To the American farm family

CR

Here's a greeting warm and tender,
We extend to you,
And a welcome, just as cordial
Is extended, too.
We should like to see you often
At the Wallace School
We, the boys and girls of Wallace,
Henry Wallace School.

We would make our lives as useful,
Kindly, good, and true,
As you've made yours, Henry Wallace,
That's what we would do.
Then we'd be to you a credit
At the Wallace School,
We, the boys and girls of Wallace,
Henry Wallace School.

—Song composed by the children of the Wallace School,
Des Moines, Iowa, at the annual celebration of the birthday of
Henry Wallace

Contents

Editor's Preface: Uncle Henry When You Need Him

I discovered the letters of the eldest Henry Wallace, the scion of the most famous farming family in American history, much too late in life, of that I am certain. When I did, the epiphany arrived with the strength of an annunciation.

Though Uncle Henry Wallace made a point of inclusiveness in writing his open letters to farm folks, espousing unusually enlightened views on child rearing and gender equity, his most poignant messages were directed at farmers' sons, sons who wore their virtue as an emblem but whose welfare, in Henry Wallace's worldview, hung perpetually in the balance.

As I read the letters for the first time, it was as if, as the saying goes, I had been "pegged." I had answered my duty as a fourth-generation Iowa farmer's son, continuing a string of one-boy families dating back to my great-grandfather, himself a Quaker farmer's son and neighbor to Jesse, Hulda, and Herbert Hoover in West Branch, Iowa. As best I could, I had defended my mother, respected my father, loved my sister, and honored my grandparents. On our family farm and woodland, I had learned reverence for all of nature's creatures. I had graduated from the top high school in my home state of Iowa, pinching my parents' pennies by agreeing to attend our state land-grant university, Iowa State University, the alma mater of Henry C. and Henry A. Wallace. At the time I thought, with all undue adolescent grandiosity, that I had the world figured out.

I attended Iowa State on scholarship, socializing myself, sometimes painfully, in the band of brothers that is an undergraduate men's dormitory. Like Henry A. Wallace, I largely kept my own company at I.S.U, avoiding fraternities and societies. I graduated with a B.A., Phi Beta Kappa, on a warm spring day in May, my name gracing the commence-

ment program with several stars of distinction, as it had in every year
since junior high. In those heady, pre-commencement days, I deferred
admission to graduate school, according to my farm father's wishes, in
favor of some "real work," and lucked into a job as a bookmobile driver
at the local public library, where, on my weekly rounds in a converted
Winnebago, I became the literary equivalent of Santa Claus.

My virtues, as I counted them then, seemed born of real courage, of
facing a progression of difficult milestones and staring them down, one
by one. I had left home in Eastern Iowa to attend college, largely paying
my own way with scholarships, manual labor in the summer, and work
at the student union bowling alley during the school year, all the while
supplementing my income pecking out stories for the *Iowa State Daily*.
All the same, like Wallace, I was restless, hastening to graduate school as
soon as my familiar beats as book-pusher and freelance journalist wore
thin.

When my grandfather took sick with cancer in my last year of graduate school, I returned home to assume the post of section editor at a
family-run newspaper in the agrarian county seat town nearest our Iowa
farm. There I learned to "write on a deadline," and "honor the reader"—
living the catchphrases otherwise uttered as mere abstractions in writers'
workshops of the sort I had attended. I completed my degree, courted,
and settled down to the comfortable routine of small-town Iowa newspaper editing. In those salad days, I remember falling into bed each night
after covering a story in a nearby small town, feeling a curious mixture of
bliss and restlessness, gratitude and ambition, and an aching sense of the
hole my grandfather's death would soon leave in my life. My grandfather, who lived only a stone's throw from my parents, had been a second
father to me.

My grampy outlived his diagnoses by two years, thank goodness,
and though the offer to take the newspaper job on a permanent basis
tempted, I hung up my pica ruler and accepted my first teaching job at a
small, Presbyterian college with a civic arts mission in the mountains of
east Tennessee. At the time, I could see with a surety bordering on smug-
ness the elegance of the design that had been laid out for me, how each
of my efforts had brought me to a place that, though far away from home,
felt right. I rented an old tobacco farm, parked my jalopy each night in
an old shed skirted by bales of hay, and woke each morning to the misty
and noble outlines of horses planted stoically in the back pasture. Had I
known of the letters of Henry Wallace then, I am sure I would have seen
fit to ignore them.

Less than four years after parading across the stage in full regalia
as the newest and youngest faculty member of my new college "family,"
I found myself weary, angry, confused, and indebted, having pulled up

roots after my second year in Tennessee for what seemed like a "bigger and better" job in Chicago, closer to the home place in Eastern Iowa. Marooned in Chicago's western suburbs, I struggled with every role I held sacred—teacher, son, mentor. More than one person I loved and respected suggested I consider stress therapy. Naturally, I refused— wouldn't any good farmer's son? I would handle it, as I always had, by myself.

Cutting my teeth on my new teaching job while pining for the farm just a few hours distant, I, mortgagee of a simple bungalow in a nondescript midwestern burg turned boomtown, became dangerously housepoor. To add insult to injury, I knew I could buy the same house for half the price back home in Iowa, where everything made better sense. In the suburbs, I found myself ill-tempered, quick to judgment, moneyaddled, and traffic-rattled. Day after day, sitting through a twelve-mile, forty-five-minute commute, I could feel my agrarian foundations eroding and my anger rising—the good life I had known on an Iowa farm more distant with each passing day and mushrooming housing development. Though I did not know it yet, I had become something like the failed sons of Uncle Henry's letters to farm boys—sons who had gone too far afield, bit off more than they could chew, and had eyes bigger than their stomachs—a thousand idioms and a thousand apologies applied. I had, in fact, become a bit like Uncle Henry, too, who endured what he called his "time of trouble" in his early middle life and who had been, even into his 40s, what Ray Stannard Baker called in *The American Magazine* a "broken down preacher of the United Presbyterian Church."

Like Henry Wallace, I wanted out.

Eventually, life, that stern taskmaster, ushered me back to my agricultural roots, as it would for Henry Wallace, though that too involved a great deal more pain and compromise than I could ever have imagined. I did find my way back home, both literally and spiritually, not as fantastically as Dorothy nor as heroically as Odysseus, but with a two-stepsbackward-one-step-forward mulishness that typifies my stubborn clan: "like a jackass in a hailstorm" we always said.

I had enough horse sense, finally, to come in out of the rain.

I began returning to the farm on weekends, responsibilities be damned. I paid more attention to the stories of my parents and grandparents, confirming the family tree, rehearsing the apocryphal family histories as well as the gilded ones. I semi-retired the hard-driven ambitions that become the double-edged sword by which generations of enterprising farmers' sons and daughters live, and die.

I studied my great-grandfather's words as if they were a Bible, long ignored, as he had penned them in his 1946 book of agrarian essays, *The Furrow and Us*, and in his Depression-era newspaper columns for the

Lisbon, Iowa *Sun*. As a teen, I had been put to sleep by talk of furrows and plows; in my thirties I stayed up late to read them. "Grandfatherism," that staple in the life of Henry A. Wallace, and the phenomenon by which interests and predilections sometimes skip a generation before re-emerging two generations hence, enthralled me. My great-grandfather, like Uncle Henry, had been an Iowa teacher, farmer, and journalist. Walter Thomas Jack had been the first to write a book to bring the author of *Plowman's Folly*—overnight agricultural darling Edward Faulkner—down to size in 1946. My Quaker great-grandfather, a man of carefully tempered passions and often sanctimonious pronouncement had, like Uncle Henry, been raised in a devout environment; like H. W., his natural curiosity and eager intellect freed him from many of his early theological demons.

In my darkest days, my great-grandfather's sentiments, on soil fertility, on conservation, on plowmanship, struck me as urgently relevant, loving yet stern. As I read his words—for the first time, truly read them—I felt, as I am sure every farmer's son and farmer's daughter must sometimes feel, that I was of him—equal to him—and yet hopelessly ill-equipped as his standard-bearer. Several passages from the book (*If you try to beat nature, nature will beat you; first the furrow, then us; the well-rounded farmer is not a wage-slave but a creative worker*) read as both whispered affirmation of my young life and unequivocal condemnation.

It was in the spirit of learning the lessons of my forebears, the truths history calls to tell us, that I first encountered the letters of the eldest Henry Wallace while browsing the agricultural history section at the University of Iowa libraries one July afternoon not long after founding and directing an Iowa summer school for agrarian youth, the School of Lost Arts. I read Wallace's *Letters to the Farm Folk* with a mixture of wonder, amusement, and incredulity; wonder because Henry Wallace was far too grand a man—the world's most famous agricultural journalist; a member of the Country Life Commission and President of the National Conservation Congress; sire of two U.S. Secretaries of Agriculture and close friend and backer of a third; pal of Theodore Roosevelt—to write with so much horse sense about the problems and pleasures of everyday farm folks; amusement because, though the open letters Wallace penned were sometimes dated in their particulars, they remained uncannily true; incredulity because the letters I now held in my hands had escaped not only my attention, but the attention of several generations of rural readers.

In a wisdom-hungry era where a search for titles beginning *Letters from...* yields hundreds of books published after the fashion of Henry Wallace—including Rilke's world-famous *Letters to a Young Poet*—I stood aghast that Uncle Henry's missives had not resurfaced. And in an

era witnessing a demographic phenomenon known as selective decentralization, wherein the census shows families once again leaving cities and suburbs for the real and imagined benefits of the country, these forgotten Henry Wallace epistles, these funny valentines, captured my heart.

For the contemporary reader, Henry Wallace is a man of alluring and illuminating contradictions—a man who underplayed his fame and fortune and refused countless overtures for highest elected office, a deeply religious man who resisted church dogma in favor of simple, profound humanism, a "man's man" of wood, field, and stream who nonetheless insisted on women's needs within and without the home, a devoted Republican with decidedly Populist sympathies. Our day knows no equally popular agrarian mavericks, though Wendell Berry and Wallace Stegner come close in the world of letters and, in politics, John McCain approximates the independent-minded fighting spirit and experience-rich ethos of the elder Wallace. Put today's Berrys, Stegners, and McCains together and you have, it is fair to say, the original Henry Wallace.

The letters in this collection have been carefully selected for their timeliness and their truth. Though farm children, and farm boys in particular, are of special concern to Uncle Henry, the adages and aphorisms here inscribed speak to every member of the farm family as well as to the urban and suburban families endeavoring to understand their country cousins. And while Emily Post and others of Wallace's era typically confine their advice to matters of etiquette and social convention, Henry Wallace rolls up his sleeves to touch on the bread and butter of rural life—how to love, play, work, agree, and disagree with uncommon grace. The letters here anthologized represent the cream of the crop from a half dozen of Henry Wallace's most popular books: *Uncle Henry's Letters to the Farm Boy* (1897); *Letters to the Farm Folk* (1915); the three-volume memoir-in-letters *Uncle Henry's Own Story of His Life* (1917-1918); and the book of posthumous letter-eulogies and testimonies, *Tributes to Henry Wallace* (1919).

While Uncle Henry's personal and business correspondence is as voluminous as it is interesting, the letters reprinted here in full for the first time in several generations are of a different sort. Written as open letters, they share two key distinctions: they are written for young people everywhere and they are emphatically public. I hope these letters bring today's readers as much joy and edification as they brought readers in Henry Wallace's day, who regarded them as a kind of Bible and child-rearing manual rolled into one. And I hope the timeless joys and challenges unique to growing up—and especially growing up on a farm—may be seen as constant across the generations. Most of all, I hope adult readers will read these letters recollecting, if not recovering, their own youth, rediscovering, as they do, the kind of parent, grandparent, or great-grandparent they aim to be. For each of these vital roles, Henry Wallace has something important to say.

Finally, I trust the multigenerational story played out in these letters both interests and affirms the power of nature and of nurture in shaping the next generation. The Romans called the mysterious and powerful spirit of family the *genius*, and they worshipped it. I hope readers share these missives, then, with their own children, grandchildren, and great-grandchildren and, in so doing, celebrate the gift that is the genius inherent in each and every family.

Zachary Michael Jack
Jones County, Iowa
June 2008

Acknowledgments

The letters to the farm boy featured in Part I originally appeared in book form in *Uncle Henry's Letters to the Farm Boy* (Wallace Publishing Company, 1897) and in several subsequent editions titled variously *Uncle Henry's Letters to the Farm Boy* (Wallace Publishing Co., 1897; MacMillan 1906, 1918); *Letters to the Farm Boy* (MacMillan 1900, 1902); and *Letters to the Farmboy* (MacMillan, 1911). For the sake of brevity, *Letters to the Farm Boy* is employed hereafter as an umbrella title and the original, 1897 volume has been used as a template. The letters to farm families that make up Part II originally appeared in book form in *Letters to Farm Folks* (Wallace Publishing Company, 1915). The missives to Uncle Henry's great-grandchildren in Part III appeared originally in book form in Wallace's three-volume memoir, *Uncle Henry's Own Story of His Life, Personal Reminiscences* (Wallace Publishing Company, 1917-1918). The Wallace tributes that constitute Part IV originally appeared in book form in *Tributes to Henry Wallace* (Wallace Publishing Company, 1919). The unattributed frontispiece photo signed "Faithfully Yours, Uncle Henry" appeared in the 1897 edition of *Uncle Henry's Letters to the Farm Boy*, while the "Sincerely Your Friend, Henry Wallace" photo appeared in *Letters to Farm Folks*. The remaining photos are drawn from the three-volume letter-memoir *Uncle Henry's Own Story of His Life*. Much of the material in this collection appeared in an earlier form in the periodical *Wallaces' Farmer* prior to its pre-1923 publication in book form. The work reprinted here is considered to be in the public domain.

Thanks, most of all, to the good-hearted "grand old man of Iowa," Uncle Henry Wallace, for daring to advise. And thanks to my own Iowa farm family for over 150 years of good horse sense and nonsense, and most especially to my great-grandparent, Walter Thomas Jack, and grandparents Edward Lee Jack and Julia Mae Jack for a lifetime of instructive letters, verse, and stories through which they showed generations of mule-headed farm kids how to live with love, grace, and good humor.

Introducing Uncle Henry Wallace—Preacher, Farmer, Editor, Philosopher, Lecturer, Counselor, Friend, Everyman

Much has been written about Henry Wallace, arguably the twentieth century's most famous rural man of letters. The bullet points of Wallace's life were well-known to the hundreds of thousands who yearly read *Wallaces' Farmer* during Uncle Henry's long tenure as editor, forming a kind of catechism for rural residents throughout the country and especially in the states where *Wallaces' Farmer* maintained its strongest readership, namely, Indiana, Illinois, Iowa, Missouri, Kansas, Nebraska, South Dakota, Minnesota, and Wisconsin.

A full biographical account of the life of the irrepressible Henry Wallace is available through a surprisingly few sources, including the Life-in-America Prize-winning biography, *The Wallaces of Iowa* (1947), penned by USDA consultant and *The Land* editor Russell Lord, Richard S. Kirkendall's *Uncle Henry, A Documentary Profile of the First Henry Wallace* (1993), and Henry Wallace's own three-volume memoir, *Uncle Henry's Own Story* (1917-1918). The two volumes of secondary scholarship cited above take a documentary approach, having the good sense to quote liberally and well Wallace's published columns, while still completely overlooking his two books of letters, one of which, *Uncle Henry's Letters to the Farm Boy*, first published by the Wallace Publishing Company in 1897, proved so popular in the early 1900s that Worldcat lists five separate editions from trade publisher MacMillan: 1900, 1902, 1906, 1911, and 1918, respectively. Unbelievably, Kirkendall's excellent and well-documented biography includes just a few indexed references to *Letters to the Farm Boy* and none to *Letters to the Farm Folk* in over two hundred pages of research. The reasons for this omission are practical and logistical (Wallace led an exceptionally long and varied life not easily exhausted in a short monograph.) and, doubtless, circumstantial—

1

Wallace's heartfelt, homespun letters purposefully offer more black earth and less ivory tower, more horse sense and less hot air.

Uncle Henry's Letters to the Farm Folk

Importantly, the thrust of this unprecedented volume is not the Henry Wallace biography, nor is it the well-documented story of *Wallaces' Farmer*. Featured here, for the first time in print in over a century, is the real heart and soul of the eldest Henry Wallace, his letters to farm families and farm children. Amid all the journalistic accolades, the cultural notoriety, and the personal friendship and favor of presidents from Teddy Roosevelt to Taft to Wilson, Wallace never lost sight of what was most important to him. In his second book of open letters, *Letters to the Farm Folk*, Wallace writes, "I have always regarded children as the all-important crop on the farm, the crop for which all other crops are grown."[1] These disarmingly simple words, spoken by the man who reared, alongside his wife, Nancy, a U.S. Secretary of Agriculture (Henry C. Wallace) and a Vice President and presidential candidate (Henry A. Wallace)—take on special relevance.

To compare Henry's Wallace's popular books of open letters to farm families to his legendary *Wallaces' Farmers* "Sabbath School Lessons" is to fully appreciate the primacy of the letters. Wallace's first book of such, *Letters to the Farm Boy,* preceded the Sunday School lessons as they appeared in *Wallaces' Farmers* and served, therefore, as a kind of advance template—a seedbed of ideas and themes. The close affinity between Uncle Henry's books of letters and his Sunday School lessons is most evident in H. W.'s second book, *Letters to the Farm Folk*, written on a wide range of subjects applicable to all members of the farm family. The books, even more perhaps than the lessons, gave Uncle Henry a truly national stage as a literary writer.

In his books for young people, Uncle Henry transcends his own religious beliefs in service to the nation's rural youth, and their parents. Editor of the *Des Moines Register and Leader* and a Wallace contemporary, Harvey Ingham describes how Henry Wallace "wrote what they [readers] wanted to read, and what came to be to them a gospel of everyday living and thinking."[2] Of these letters and lessons, Ingham continues, Uncle Henry made a "correspondence school for the farmers of the great middle-west, in the principles of right living, right thinking, right farming.... As editor, he renewed his youth at the fountain of general optimism that kept him young in spirit and fresh in the hearts of his readers."[3] After Wallace's death in 1916, Ingham's *Register* printed a story describing the teary-eyed gathering of the Des Moines YMCA, to which Wallace had been both supporter and benefactor. Addressing Wallace's devotion to young people as expressed in Uncle Henry's books to farm boys and farm folks, H. W. Byers said, "Upon every question affecting the welfare of the young men and young women of this country, he [Wallace] took a decided stand, always for what he firmly believed to be the right."[4] F. W. Beckman, a professor of agricultural journalism at the Iowa Agricultural College, now Iowa State University, wrote in the *Iowa Agriculturalist*, "His [Wallace's] letters to the farm boys, published in recent years, were so full of sympathy and wisdom that from every direction came the request they be put in book form. That volume deserves a place in every library."[5] The same remains true today.

Unusual in his day, Henry Wallace considered the health and welfare of the farm family as a whole, and reached out to the long-ignored farm mother. Long-time friend Tama Jim Wilson, the man who Uncle Henry helped become Secretary of Agriculture for McKinley, Roosevelt, and Taft, wrote, "He [Wallace] has been in sympathy with the farmer, the farmer's wife, the farmer's boy, the farmer's girl, and with the hired man."[6] In his second book of letters, *Letters to the Farm Folk*, Uncle Henry's empathies grow still more inclusive, resting especially with the farmer's wife, just as they had in his groundbreaking work on the County Life Commission. "Of all farm folks," Uncle Henry asserts in his *Letters to the Farm Folk*, "the mother on the farm is the most important; in fact, she is indispensable."[7] He goes on to gently remind the farmer of the obvious, namely that he ought to show as much care and concern for the welfare of his wife and children as for his farm. Henry Wallace, the product of a mother who had raised eight children to adulthood on the farm, used his good offices and influence over the farmer to advocate on behalf of farm women's material and emotional needs. In one particularly heartfelt passage, Uncle Henry writes to fellow farmers, man to man:

> Another thing you can do: You can manifest to your wife something of your affection which you have lavished upon her in courting days. I say, "manifest." There is a great deal of affection among men towards their wives that remains unspoken. Men forget to continue to play the lover, and assume that having once been convinced them of their undying affection, there is no need of saying nice things after the honeymoon is past. Women are not built that way. They hunger for words of affection and endearment and commendation, and hunger even more for the affection that flows from tenderness of heart, and all the more because it is an expression of the tenderness of the strong.[8]

In fact, many who read Henry Wallace's inimitable letters for the first time in *Letters to the Farm Boy* and *Letters to the Farm Folk*, considered them the purest distillation of his voluminous life's work. As early as 1916, Iowa attorney, railroad commissioner, and former U.S. Senate candidate Clifford Thorne made the argument that several Wallace biographers and scholars since have all too readily forgotten: "His [Wallace's] books," Thorne reminded the Des Moines YMCA, "constitute one of the principal portions of his life work." Indeed Thorne and other Wallace contemporaries who experienced the popularity of Uncle Henry's epistles firsthand knew something subsequent generations of readers have missed—that while the avid, mostly male readers of the magazine *Wallaces' Farmer* were largely concentrated in the Midwest and Plains, *Uncle Henry's Letters to the Farm Boy* and *Letters to the Farm Folk* reached men, women, and children throughout the land, finding them, literally and figuratively, where they lived.

It was this universal yet home-grounded ethos, best achieved in his letters to farm folks, that earned Wallace the affectionate sobriquet "Uncle" and made of his last five years on earth a kind of national victory lap. In nominating the nearly seventy-five-year-old Wallace for the presidency of the Conservation Congress in 1910, nominators tabbed "a man in whom there is no guile, who is not only well-known in this country but who has international fame."[9] Seconding the motion was none other than Gifford Pinchot, considered by many to be the father of the conservation movement in addition to being Teddy Roosevelt's chief forester and right-hand man. Addressing Wallace's avuncular moniker in extemporaneous testimony, Pinchot begins:

> I pray for your indulgence for a moment while I try to say a little of what I think about "Uncle Henry" Wallace. I call him "Uncle Henry" for the best of all possible reasons—that when a man has reached his age in a life of usefulness, he becomes, in a sense, the forebear of the rest of us, and our affectionate esteem naturally expresses itself in calling him "Uncle"; and I say "Uncle Henry" Wallace because I love him.[10]

That same year, 1910, the Wallace hit parade continued with a profile by one of America's leading journalists, editor of *The American Magazine*, Ray Stannard Baker, who, in a feature titled "Interesting People," praised *Letters to the Farm Boy* for its singular nature and described its author as "a sort of oracle for advice on everything from the best ways of feeding... calves to bringing up boys."[11] Moreover, Baker claimed, Henry Wallace had helped "lay the foundations for the present progressive political movement" as represented by Teddy Roosevelt and Woodrow Wilson, among others. *Letters to the Farm Boy*, Baker declares, "has had a wide sale," a phenomenon he attributes to the letters' "understanding of the deepest life of the American farmer, for sound sense, for ripe philosophy, for pungency of the English in which they are written." The epistles, he concludes, "are unique among American books."[12]

In hindsight, Wallace offers contemporary readers a strikingly rich and unusually varied legacy, one that extends well beyond agricultural and environmental studies where his name long ago entered the pantheon. In the literary world, Wallace's advice columns made the Midwest ever after synonymous with sound advice of the kind later doled out by Sioux City, Iowa natives and identical twins Ann Landers and Abigail Van Buren. As a cultural critic, Wallace's anticommercial, profamily, pro-education stance prefigured what is, to the present day, quintessential Middle American politics. In accepting the presidency of the National Conservation Congress in 1910, Wallace articulated a still relevant, strikingly contemporary vision for America, saying, "I believe that the mission of this nation is not to build great cities, not to be a world power, not to amass wealth untold, but to develop character."[13]

Where have you gone, Henry Wallace?

Given Uncle Henry's half million weekly readers, his presidency of the National Conservation Congress, his appointment to the Country Life Commission, and his widely acknowledged status as the most influential agrarian of his era, what impresses is not the Henry Wallace-dedicated biographies and articles extant, but the conspicuous lack thereof. After all, Federal Farm Loan Bureau chief, editor of *Country Gentleman*, and popular fiction writer Herbert Quick once claimed that Henry Wallace "would be remembered by the farmers and many others when the great mass of governors, senators, congressmen, justices of the supreme court and cabinet officers of the day are forgotten, for he worked with the people, not over them."[14] In one sense, Quick's words have proven prophetic, as Henry Wallace's name has indeed outlasted Tama Jim Wilson's, the man Wallace made Congressman and Secretary of Agriculture, or Iowa Governor Larrabee, who also owed his office, in large part, to the Henry Wallace publicity machine. But in another sense, it is the name *Henry Wallace*, and the ethos it represents, that has endured, much more than any single agricultural, environmental, or theological accomplishment among many achieved by the "grand old man of Iowa." By contrast, Liberty Hyde Bailey, who, alone with Wallace, could reasonably lay claim to the title of the era's foremost agrarian, is studied in the halls

of our nation's college and universities for his pioneering work in nature study, conservation, horticulture, and agriculture. Yet Bailey enjoys a fraction of the abiding star power or name recognition of his friend and colleague. In part, this difference might be chalked up to personality (Wallace's folksy charm versus Bailey's organizationally minded studiousness), publication type (popular writing versus scholarship), or even to region (the "new frontier" of Wallace's Midwest versus Bailey's "old-news" East).

Given Uncle Henry's historical popularity, the enduring power of his name, and the zeitgeist of what it stood for—Good Farming, Clear Thinking, Right Living—the dearth of scholarly material on Uncle Henry Wallace should give us pause. Several explanations suggest themselves. First and foremost, in writing an exhaustive, three-volume memoir shortly before his death, Uncle Henry beat all biographers to the documentary punch. Second, Uncle Henry's *Wallaces' Farmer* aside, Wallace's rise to national prominence preceded the widespread rise of the popular farm and conservation press that might otherwise have painstakingly documented his ascendance. Scholars have, therefore, had to make due with a relative few general interest magazines in wide circulation in the late nineteenth and early twentieth century and still available on microfilm and in hard copy—publications such as *The American Magazine*, *American Mercury*, and *The World's Work*, all of which covered Wallace alongside other cultural movers and shakers. A third possible reason for the widespread neglect of the eldest Henry Wallace is familial. Wallace's son and grandson, both of whom would serve, in close succession, as United States Secretary of Agriculture, equaled Uncle Henry's own considerable fame. Had grandson Henry A. Wallace not achieved worldwide fame as a Vice President, candidate for President, and father of the seed corn industry, his grandfather might well have garnered the undivided attention of historians, scholars, and lovers of the land.

Popular history everywhere has conflated the three prominent Henry Wallaces—Uncle Henry, Henry C. and Henry A.—making them one in the public mind, in much the same way that history has blurred the distinctions between father-son presidents John Adams and John Quincy Adams, and as it will one day do with George Bush, Sr. and George W. Bush. Finally, the full impact of Uncle Henry's potential fame was mitigated by historical happenstance. Just as Henry Wallace's star reached its zenith with the publication of his second book of letters, *Letters to the Farm Folk*, in 1915, the urgent industrial needs of World War I drew the country's attention away from such "abstractions" as character education and good child-rearing practices. After World War I, a war Uncle Henry ardently resisted, agricultural prices and markets once again destabilized

and, for the most part, remained that way. As if preserved in amber, Uncle Henry's timeless advice from the Golden Age of agriculture was left for future readers to unearth.

While his sons were newsmakers, Uncle Henry was, quite happily, a culture-maker, a true national celebrity in the days before newspapers and radio broadcasts reached rural audiences in significant numbers. Oral histories and memoirs of the period well document the unprecedented influence Henry Wallace had in rural communities throughout middle America. Wallace's intimate expositions, inclusive of both his book-length open letters and *Wallaces' Farmer* Sabbath School Lessons, formed a Bible, quite literally, for country folk like Homer Croy, who in his memoir *Country Cured*, fondly remembers nodding off to sleep each night while reading and rereading *Wallaces' Farmer*. "We also took the county weekly," Croy writes, "but it wasn't the world of enchantment that *Wallaces' Farmer* was.[15] As a phrase, "land of enchantment" perfectly captures Uncle Henry's Oz-like rhetorical powers, his real-life wizardry.

Henry Wallace's project was cultural as well as familial. He made little distinction between exercising his good offices within his clan and without, as he treated his half million readers just like family. When grandson Henry A. Wallace was born in 1888, Uncle Henry, according to Wallace scholar Richard Kirkendall, "made a conscious, deliberate effort over the course of a quarter century to shape the boy and the young man."[16] The same could be said of Uncle Henry's influence over all the rural boys of his son's and grandson's generation, for whom he was a kind of surrogate father, confessor, and friend. Millburn Lincoln Wilson, Franklin D. Roosevelt adviser, national director of the Extension Service, and native Iowan, read *Letters to a Farm Boy* as a lad in the 1890s and never forgot it. "The Wallaces helped form my mind and character," Wilson reflected, "just as they did for hundreds of thousands of other folks."[17] The Elder Henry Wallace never veiled his paternal influence; unlike Oz, Wallace wanted his readers to pay attention to both the man and the machine. A prime example is his 1902 editorial entitled "A Word to the Bright Boy's Father." Addressing fathers of boys in the corn belt, H. W. offered full transparency, writing, "Now we want to help you to keep these boys on the farm and to do it by making them better farmers.... They can do it if they get on the right track. We want to put them on the right track. We want you to help us. We want to do it by teaching them how to grow to the best advantage the principal crop of the Mississippi Valley—corn."[18]

The vagaries of history may be to blame for the relative shortage of biography and scholarship on Uncle Henry Wallace, yet those same

machinations promise cyclical resurgence. Nearing the one-hundred-year anniversary of Uncle Henry's death, the public mind is just now returning to the idea of the agrarian as crucial cultural steward and foremost social critic. The leading edge of this cultural moment, coming at a time when census data shows country living once again in fashion, is evident politically in the popular, earth-centered politics of Al Gore and in the disproportionate national interest in the successful election of John Tester, a Montana organic farmer, to the United States Senate in 2006.

Culturally, Henry Wallace's high regard for the environment and for human life from cradle to grave is alive and well in the form of New York farmer-physician-activist-author Dr. William "Bill" Thomas, whose 258-acre farm serves as a template for a revolution in care for the aged: "eldershire."[19] Taking a page from Henry Wallace's playbook, Thomas's most popular book, *In the Arms of Elders*, casts the next generation, Thomas's daughters, in the starring role of a multigenerational, farm-grounded passion play. Elsewhere, American film and television screens featured, in 2005, a twenty-first-century version of the maverick midwestern farmer in *The Real Dirt on Farmer John*, the award-winning Taggert Siegel documentary of "failed" third-generation Illinois farmer John Peterson who returns to farming, writing, and speaking after discovering—as Wallace had more than a century earlier—agriculture's true roots in sustainable community.

In a cultural moment when studies allege alarming and widespread deficiencies in America's children, and particularly in American boys, Uncle Henry's *Letters to the Farm Boy* and *Letters to the Farm Folk* merit a dusting off. Henry Wallace Sr.'s arguments for the benefits of rural living and the necessity of providing children with ennobling work to do find contemporary ground in the plethora of Outward Bound programs, wilderness retreats, and dude ranches for American youth. Even the strong viewership of television programs such as Public Broadcasting's *Texas Ranch House*, where youngsters are transported back in time to the early Henry Wallace era to live as pioneer ranchers, belies our collective sense that old-time, rural wisdoms need reliving. The description for the show trumpets the drama inherent in an instructive historical moment, as it advertises a "brave family and a diverse group of cowboys-at-heart [who] discover how myth...meshes with reality—and what the saddle-sore, rope-burned, and sun-blistered ranch life really was like."[20]

Science, too, increasingly recommends for our children a timeless, hands-on experience of the kind Wallace provided his son Harry and his grandson Henry on the family land and in the family business. A 2003 Cornell University study, one of many of its kind, suggests that those children who merely grow up with a "green view" from their window are better equipped to handle stress.[21] Similarly, nature study, reports the California Department of Education, resulted in twenty-seven percent better science scores for those studying in the outdoors when compared to those trapped in traditional classrooms.[22] And yet for all their newness, the studies confirm what Henry Wallace already knew. Just a few short months after Uncle Henry's death in 1916, *The American Review of Reviews* offered a digest of the testimonial offered by editor Herbert Quick in the pages of his magazine, *Country Gentleman*. Quick offers this synopsis of Wallace's agrarian and environmental argument: the "passing down to future generations a well-kept farm, unimpaired in fertility, and adapted to the nourishment of a happy, wholesome life is in itself an act of worship."[23] In closing, Quick asks rhetorically whether anyone can fully appreciate the "sweetening and uplifting [of] our national life"[24] made possible by Henry Wallace.

The Education of Henry Wallace

The details of Uncle Henry's own life, oft-cited as anecdotes in support of his singular and universal popularity during the Golden Age of American agriculture, approach the level of folkloric parable on a near-par with Washington and his cherry tree or Lincoln and his log cabin. For those who know well the history of the first Henry Wallace, the comparison between the popularity of Washington and Lincoln and Wallace is less an exaggeration than it might first appear.

If numerological theories appeal, it is worth noting that Henry Wallace died on Washington's birthday and that his grandson, Henry A. Wallace, was called to serve as Secretary of Agriculture on Lincoln's birthday. And if the numerical evidence seems too far-out or far-fetched, one need only examine the literature. Scarcely four pages into his definitive 1947 biography, *The Wallaces of Iowa*, Russell Lord writes that the first Henry Wallace "would have made a splendid companion for George Washington had their days on earth coincided."[25] "The present head of the family, Henry Agard Wallace," notes Lord, "is of a mind and character that will delight Jefferson in heaven."[26]

In the hundreds of letters and testimonials collected under the title of *Tributes to Henry Wallace* and published by the Wallace Publishing company after Uncle Henry's passing on February 22, 1916, Lincoln is repeatedly conjured as Henry Wallace analog. In comparing Uncle Henry Wallace to Lincoln, W. P. Johnston, a college classmate of Wallace, writes, "Mr. Lincoln's life becomes interesting not only for the things that he did, and the words that he uttered, but because of the influences

about his childhood and boyhood that gave him his bent of mind and his stability of character."[27] Wallace, Johnston suggests, is the same type of man. Though far more modest than Lincoln's, Wallace's funeral was itself a thoroughly public event attended by hundreds if not thousands; the *Des Moines Capital* recorded "a steady stream of sorrowing humanity....Gray-haired men mingled with school children; mothers bent with age, high school girls, ministers, businessmen, students, clerks, workmen from factories and shops, newsboys with hushed voices, and persons from every walk of life—all friends of 'Uncle Henry'—joined in paying this last token of respect and reverence."[28] Similarly, the *Des Moines Evening Tribune* documented attendees of "every class." "Bankers," the *Tribune* noted, "walked with laborers. Girls from the Young Women's Christian Association stopped to look at their benefactor, and boys from the Young Men's Christian Association joined them in grief." The report of the day's events by the *Des Moines Register and Leader*, the third of the then-great trio of Des Monies newspapers, painted a picture of funeral-goers of "all creeds, all faiths, all political beliefs, and all positions in life."[29] That a young Henry Wallace would, as a seminarian, have occasion to dine with his uncle and Abraham Lincoln and later cross paths with the Great Emancipator while Wallace served as a chaplain for the Army of the James seemed preordained. Calling their dinner together "the event that made the deepest impression on me," Wallace remembers a remarkable Lincoln—"tall, awkward, ungainly, homely—a man who paid little attention to dress, and had what seemed to be the saddest face and saddest eyes that I ever saw in a human being."[30] That Wallace was himself over six feet tall, thin as a rail, and burdened by ill health and heavy responsibility throughout his life may partially explain the affinity.

The Henry Wallace story is a thoroughly American one, accounting, in part, for the mass appeal of a man leading rural historian R. Douglas Hurt calls "the preeminent advocate for farmers during the late nineteenth and early twentieth century" as well as a "unifying voice urging improvements through science, technology, and education."[31] Henry Wallace, at once dark horse and golden child, was the only one of nine children of John and Martha Wallace to live past the age of 30. A Pennsylvania farmer's son, Wallace was a late bloomer, a trying child, impatient with farm chores, hasty in his work, and quick to temper. Afforded membership in a preparatory boys academy via family ties rather than wealth or position, Wallace was still raw, still green, well into his teenage years at the Geneva Hall school, where he recalled himself as an "awkward boy"[32] and where even his professor's wife chastised him for his disheveled appearance. Revelatory of Wallace's lifelong love of the underdog, such upbraidings became his favorite anecdotes, especially one particular response his fashion faux pas evoked from his professor, who, after politely listening to his wife's critique of the slovenly Wallace, said to the boy, "She is giving you good advice; but I would advise you not to neglect your studies to take on polish. If you master your studies, there will be something to polish, and that will come by and by."[33]

The Henry Wallace who attended Jefferson College several years later was not much more refined, as he remained prone to gluttony, occasional foolishness, and the usual adolescent awkwardness. Wallace was turning into an unlikely leader, though, acting as judge and juror in moot courts whenever a dispute flared among his peers. And increasingly he was being taken under the wing of professors who seemed almost incredulous that they had agreed to mentor such a rough charge. Wallace's teachers got through to him by keeping it simple, as was the case of Dr. Joseph Alden, who told him, "Wallace, there are some men who can think and can not talk; and there are other men who can talk but can not think. I want you to learn to do both."[34] And learn he did.

Schooling at Jefferson College in the days leading up to the Civil War was not, it should be noted, anything like the education received by the nation's fortunate sons in contemporary academies. Jefferson College featured none of the back-slapping chumminess or silver-spooned fraternity of a Yale or a Harvard. In his memoirs, Henry Wallace colors it thusly:

> There were in Jefferson College at the time of which I write no games or sports, no baseball nor football, with their factions and leaders, nor athletic contests with other colleges, involving large expense of time and money. I do not even remember if we had a college yell. If we did, I have forgotten it....In short, we were gentlemen, coming mainly from farm homes where money was none too plentiful, and with sincere and earnest purpose of fitting ourselves for the serious business of life.[35]

The native seriousness of which Wallace writes, coupled with the hopes of his pious mother, would eventually lead him west to the Illinois-Iowa border to seminary at Monmouth, Illinois, but not before the contrarianism and independent thinking that had become Wallace's trademark surfaced at United Presbyterian Theological Seminary at Allegheny, where biographer Russell Lord writes that the free-thinking Wallace was "distressed and irresolute, a young man torn from his faith in mixed spiritual moorings, restless and sore at heart...fitting less and less readily

into the routine indoctrination there in practice."[36] That Wallace resisted "routine indoctrination," and that he was something less—or something more—than a tradition-bound, Presbyterian minister-in-training—would ultimately benefit the many farm children and farm folks who read his letters and who found in Wallace a voice that was empathetic rather than holier-than-thou. Wallace's recognition of his difference is recorded in the second volume of his three-part memoir, *Uncle Henry's Own Story*, where he remembers: "The question now arose with me whether I should go back to Allegheny or west to Monmouth, Illinois. I had seen enough of the rigid opinions of the leaders of the United Presbyterian Church in Pennsylvania to convince me that I would never be an acceptable preacher in that part of the country."[37] As it turned out, Wallace, despite being a gifted sermonizer and a willing official at weddings and funerals throughout his life, would not find the fulfillment he sought until he gave up his congregation years later and settled into farming, writing, and editing. When blended with his theological training, these eclectic talents would make him the most beloved agrarian of his day.

Ordained in 1863, Wallace was supplied by the seminary at Monmouth to two small churches on opposite sides of the Mississippi: one in Rock Island, Illinois and the other in Davenport, Iowa—the river-divided cities symbolic of Wallace's own riven nature. In his first years in the pulpit, Wallace took a middle road on most issues, hoping not to privilege either of his divergent congregations, both of whom suspected him of playing favorites. His religious philosophy rankled, as did his habit of smoking stogies and provoking congregation members. For instance, though the Presbyterian church was dead set against dancing, especially dancing by the young, Wallace was willing to defer to youthful inclinations, though he did not like to dance himself. When the church organ became controversial, Wallace stood by a preacher's right to have an organ played in the church, though he never chose to augment his own services with one. Showing a glimpse of the rabble-rousing, populist indignation that would distinguish his later newspaper writing and editing, Wallace spoke out vehemently against slavery to congregations sprinkled with Southern sympathizers while preaching against graft where he found it in local army camps and among Davenport, Iowa businessmen. Wallace's justice-minded muckraking from the lectern pleased some while alienating others, including, especially, the wealthier members of Wallace's fold. During the Rock Island and Davenport years, Wallace developed the social consciousness that would endear him to Teddy Roosevelt and the rest of the Progressives: he raised funds for local libraries, created a "poor man's club" with readings and refreshments designed to compete with saloons, and inaugurated a lecture course that was held—appropri-

ately enough given Wallace's farm roots—in a stable.[38] Wallace's identity as an agrarian populist took root in 1862 and 1863, when he began to braid religion, oratory, politics, and social justice in a single charismatic package. Russell Lord notes of Wallace's developing inclinations: "In the expository style of preaching that he was developing, Wallace gave preference to a conversational tone. The orthodox high-boxed pulpit of his Davenport Church irked him."[39]

Wallace's unorthodox ministering along the Illinois-Iowa border continued against a backdrop of confrontations, antagonisms, and tragedies leavened only by his marriage to Nancy Ann Cantrell in 1863 and the birth of Henry Cantrell Wallace—the future father of world-famous Henry A. Wallace—in 1866. In between, Wallace realized death as a constant companion, most horrifically as a Civil War army chaplain serving under the auspices of the nondenominational Christian Commission. Death and disease affected Wallace, too, on the Iowa-Illinois home front, where news reached him of the death of his brother James from consumption—the first of seven brothers and sisters to die from the same disease.[40] If the tragedies afflicting Henry Wallace's blood kin were not enough, battlefield deaths suffered within the family of his in-laws, the Cantwells, mounted. Nancy Wallace, pregnant with Harry C., took Lincoln's death as an omen of what she imagined would befall her husband on his long journey back from his army chaplaincy in Virginia. In his memoirs, Henry Wallace recalls: "When that morning she saw in the papers that Lincoln had been killed, she ran down to my brother-in-law's store, crying, and saying that the Rebels had killed Lincoln, that they were going to kill all great men, and Mr. Wallace would be the next one."[41] The fact that "Mr. Wallace" did make it home offered a rare moment of celebration for the young couple, who, according to Russell Lord, endured the death of Wallace's favorite sister in 1865, the death of the second eldest brother Ross in 1868, and the indignity of the forced sale of Wallace's boyhood farm home in Pennsylvania—Spring Mount—in 1869. Meanwhile, Wallace's mother and father were forced to abandon the Pennsylvania home place to rent a house in Rock Island, Illinois with Henry Wallace's three surviving siblings. Preaching two sermons on Sundays while caregiving for his ailing clan, Henry Wallace's health likewise faltered. His life, at once hard-pressed and blessed, increasingly assumed the character of the great American meta-narrative: a story of birth, death, marriage, crises, and homecoming that George Bailey, a dead ringer for Wallace in many ways, would make famous thirty years after Wallace's death in the Frank Capra classic *It's a Wonderful Life*.

Importantly, it was the newfound proximity of Wallace's father, the old farmer, that precipitated the next stage in Uncle Henry's ultimate re-

turn to the land. Financed by a thousand-dollar loan from John Wallace, Reverend Henry Wallace purchased a plot of land in Scott County, Iowa, before buying and selling more land downriver in Burlington, Iowa and further west in Adair County. With each successive sale, he turned such tidy profits that even his cautious father directed him to invest money on his behalf in farm ground in Adair County. Meanwhile, Reverend Wallace's professional life was transforming, as he resigned his position in what is now the Quad Cities in the winter of 1870-1871 and accepted a "country call"[42] in the town of Morning Sun, Iowa, where he ministered in lay clothing to farmers in the fields. The change from urban to rural precipitated a change in Wallace's communication style, making his speech plainer, more open. In the chapter of his memoir entitled "Pastoral Visitation," Wallace recalls the simple reciprocity of his communications with local farmers. "I talked to these people freely about their farms and farming, sat at their feet, so to speak, and learned enough farming to ask them intelligent questions," Wallace wrote. "Here I got my first idea that if a man wants to preach effectively to country people, he must do it in terms of farm life."[43] In his six-year tenure at Morning Sun, the increasing stability of the Wallace clan was interrupted by more deaths—Henry's brother Daniel in the winter of 1871–1872, his father in 1873, his sister Margaret in 1876, and his brother John, the last of the family, in the fall of 1876. Meanwhile, Henry Wallace's own health improved with the birth of his second son, John, in 1871 and as a result of increased time spent outdoors hunting and beekeeping. With his characteristic wry humor, Wallace would later recall with wonder that the staid Presbyterian congregation of Morning Sun put up with a preacher who did "unministerial things" such as hunting. But, he added, "My people endured it."[44]

The year 1875, one of his last in Morning Sun, proves a particularly crucial chapter in the life of Uncle Henry, as the young reverend, resistant to politics his entire life, was nominated against his will for a position in the Iowa Senate. In his memoirs, Wallace's disgust with himself for compromising his principles proves palpable. He remembers being fed "the usual dope"[45] by the state committee who supplied him with campaign rhetoric that he, in turn, relayed, absent both passion and belief, to the electorate. In a nutshell, the election was a contest between the rival towns of Wapello and Columbus Junction, both of whom were vying to be the county seat of Louisa County. When the Columbus Junction contingent arrived at his door to remind Wallace that he ought to take a side or else, Wallace is alleged to have said, "Then I will be beaten, for I have given my word that I will take no part in this controversy, and I won't do it."[46] By his own count, Wallace lost the election by twenty-three votes, insisting that he was "sincerely glad" to have been defeated, though cha-grined that he had disappointed his wife's political ambitions for him.

Wallace the Obstinate, Wallace the Purist, and Wallace the Public Advocate represent three aspects of the complicated personality Wallace conjures for the balance of a life in which he assiduously avoids running for public office while still managing to represent the farmers' political interests. Though he would never again mount a campaign, Wallace's pro-digious influence in the Progressive period greased the wheels of Iowa's "Tama" Jim Wilson's appointment as the first Secretary of Agriculture, though Uncle Henry himself resisted "sure bet" calls for a Wallace cam-paign for both Iowa governor and U.S. Senator. True to form, Henry Wallace would have none of it, preferring instead to cultivate his sublime amateurism and third-party objectivity.

Henry Wallace: Editor, Newspaperman, and Advice-Giver

In 1877, a 41-year-old Uncle Henry resigned his ministry in Morning Sun under the orders of his doctor, who was said to have told the middle-aged reverend that it was "either the pulpit or the boneyard."[47] Nancy Wallace, who had watched all of her husband's immediate kin pass on, worried not only for her husband but for her growing family of four children—sons Harry and John and daughters Josephine and Harriet—all the time in danger of losing their father.

Wallace's accounting of his difficult leave-taking from Morning Sun further dramatizes his legend, as Wallace would be given up for dead, quite literally, before his resurrection as a central Iowa clover farm-er and newspaper columnist. Wallace recalls that the westbound train he took from Burlington, Iowa was full of preachers who regarded him practically as a "dead man."[48] According to his memoirs, Wallace was so disgusted at their morbidity that he sat on the steps of the train car all the way from Keokuk, Iowa—though not before setting them straight. The account is vintage Wallace: "To the most solemn of the whole lot, I said, 'You needn't select a grave or employ an undertaker. You needn't even appoint a committee of condolence to my widow: for I shall not die.'"[49]

Indeed, Wallace's trek to the fertile grounds of Madison County, the very same Madison County later made synonymous with romantic resurrection in Robert James Waller's bestselling *Bridges of Madison County*, proved to be just what the doctor ordered for Wallace, whose health improved on arrival. The same could not be said for his financial condition, which worsened in the aftermath of unprofitable experiments

in cattle breeding and a failed launch of a creamery business. It was either a sign of Wallace's continued hard luck or evidence of some special, existential trial set for him that he entered farming at exactly the wrong time—years in the late 1870s remembered for their wild fluctuations and rampant speculation. Overplowing and overcapitalizing the land alongside his farming peers, Wallace learned his lesson quickly, adopting a conservationist mindset that would, years later, result in his presidency of the National Conservation Congress in 1910. In *Clover Culture*, a book published by the Homestead Company in 1892, Wallace writes in the vanguard of the movement, juxtaposing "soil robbers" with "intelligent farmers" who allow their lands to rejuvenate under a nitrogen-rich cover of grasses.[50] Turning to clover as an alternative to exclusive grain growing on impoverished soils, Wallace admits that he overlooked the fundamentals in mismanaging his father's estate and his own, recollecting his tendency to "begin at the wrong end" by focusing on the breeding of fine livestock rather than the foundational "study of soils and methods of cultivation."[51]

But it is these hard times, and Wallace's gradual frittering away of his father's inheritance, that, in this classic redemption narrative, provide the springboard for Uncle Henry's true calling as an agricultural journalist and mover of men. Asked to make the main speech at the Fourth of July festivities in Winterset, Iowa, Wallace, who was then attending to his gravely ill son, Ross, took a friend's advice and decided to deliver a speech of substance rather than easy humor. Though the speech, which launched Wallace's famed, late-life career, amounted to a populist rant against corrupt politicians, ineffectual, city-bred teachers, and quack doctors foisting false cures on rural patients, it was viewed as highly partisan. The Democratic paper in Winterset, the *Beacon Light*, condemned him, Wallace remembers, as a "discredited minister of the gospel who happened to have a little money and was pretending to teach agriculture to the farmers,"[52] causing the Republican paper, the *Madisonian* to come to Wallace's defense both in print and in person. The editor of the *Madisonian*, in fact, offered Wallace a job as an agricultural section editor, which Wallace accepted on the condition that he have, as he called it "absolute control of the page."[53] For several months thereafter, Wallace filled his page with conservationist warnings about declining soil fertility, about the need for grass, about what, in contemporary parlance, is called "value-added" agriculture. But when, roughly ninety days into his new post as an agricultural columnist, his editor attempted to censor Wallace's charges that politicians were insincere in their representation of farming interests, Wallace resisted and, as a result of his obstinacy, was sent packing.

That Henry Wallace could no more hold down a job than bite his tongue is, once more, part of his mythos—a mixture of advocacy, irascibility, and insubordination that endeared him to rebels and iconoclasts everywhere. Some twenty years later, when a gray-bearded Wallace penned his *Uncle Henry's Letters to the Farm Boy*, his past worked to underscore his endorsement of gumption, enterprise, self-reliance, and conscience. Ironically, his own history of overreaction legitimized his advice to young men to control their tempers and to curb their desire to make money hard and fast, fist over fist—both vices to which the younger Wallace fell victim.

Rather than be a yes-man to a conservative boss, Wallace, after his dismissal, bought a half interest in another newspaper in the county and, when he realized that his editor-partner was as new to the industry as he was, bought out his partner.[54] The same Henry Wallace who would, in death, be praised as a consummate, concerned boss with a knack for delegating duties and affirming his employees, was, well into his early 40s, a loose cannon and a firebrand, eager for influence and justice. Wallace's party-independent *Chronicle* took on all comers, as H. W. regained his form as a lay minister, translating his passions into journalistic commentary and tripling the subscription base by his own calculations. Once Wallace had positioned his newspaper as the moderate alternative to the partisan papers in town, he managed to attract still more readers who had defected from the bickering newsrags. Within a few years, Wallace had bought out the rival Republican paper in Winterset, the *Madisonian*, using it as a mouthpiece for the campaign of his friend "Tama" Jim Wilson for Congress and a chance to "give vent to his son Harry's surplus energy"[55] by giving the young lad a pressroom in which to cut his teeth.

The father-and-son, Henry and Harry, consolidation of the *Madisonian* marked yet another turning point in the career of Henry Wallace, as in Harry (Henry C.) Wallace, the "old man" had found a partner in what had become the family business. After the custom of agrarian families, the fortunes of father and son would be inextricably linked; when Harry went off to Iowa State Agricultural College, for example, and complained to his father of poor instruction by professors with little practical farm experience, Henry and Harry succeeded in placing Henry's friend Seaman Knapp into the presidency of the College; within three years of assuming the post, Knapp made good by drafting a landmark piece of agricultural legislation: The Hatch Act. Likewise, when Harry returned home after his sophomore year at college to find that one of his father's farm tenants had pulled up stakes, the elder Wallace agreed to allow his son to rent the farm, a scenario of recommended trial ownership that emerges several times in Uncle Henry's *Letters to the Farm Boy*.

Dark Days, *Wallaces' Farmer*, and the advent of "Uncle Henry"

Though Henry Wallace was a proud father-in-law and a soon-to-be-grandfather, he was not yet, in the late 1880s, the wise, fully realized soul of *Letters to the Farm Boy* and *Letters to the Farm Folk*. Instead, he was a marginally successful businessman, impulsive and restless. His desire to help the farmer led him first to form the Farmers' Protective Cooperative and, by way of the cooperative, to a profoundly antimonopoly, antirailroad stance that would, in time, result in yet another firing. By this point Wallace's influence in Iowa was sufficient to turn both gubernatorial and congressional elections. Any candidate for public office in Iowa had to answer, directly or indirectly, to Henry Wallace.

As Uncle Henry approached his fiftieth birthday, he continued, as the saying goes, to burn the candle at both ends, combining the idealism of his seminary days with hard-driven business acumen. He was now owner of two newspapers and, fatefully, would become the editor of a third, the *Iowa Homestead*, by request of new owner J. H. Duffus. Wallace took the position in part as a favor to Duffus, a Wallace acquaintance wanting to give his new publication a good start, and because all it required of Uncle Henry was a single page of agricultural copy each week.[56] When the *Iowa Homestead* changed hands a relatively short time later, this time falling into the hands of a perfect stranger, J. M. Pierce, Henry Wallace struck what he thought was a gentleman's deal with his new boss. He offered to continue editing the paper in return for a mere ten dollars a week, asking only that he be given a stock option if the paper succeeded.[57]

Wallace's assumption of the editorship of the *Homestead* occurred in the middle of a developing agricultural depression and mounting agrarian frustration at money-lenders, middlemen, and, especially, at the railroads and their exorbitant rates. And though he disagreed with the vehemence and even evangelism of farmer organizations like the Grange and the Farmer's Alliance, Wallace sided with Populists against the railroads and the politicians who habitually kowtowed to their demands. Characteristically nonpartisan, Wallace's *Homestead* managed to defeat, by Wallace's own count, seven railroad-bought Republicans and six Democrats in the 1887 elections and managed, with the help of Tama Jim Wilson in the House of Representatives, to advocate for the establishment of the Interstate Commerce Commission. By 1888, Wallace became, notes biographer Russell Lord, "a new sort of state political boss, without a machine, with scant funds, but with a mixed and trusting following of wild-eyed agrarians and conservative farmers."[58]

In an age when more than half of Americans in many regions of the country still made their living from the farm, Henry Wallace had found a cause, an era, and a medium perfectly suited to his passions. The discontent of farm advocates had reached a fevered pitch. Antiestablishment, antimonopoly speeches turned both venomous and violent, especially in the South, where Tom Watson stirred the flames with this call to arms: "Men of the Country, let the fire of this revolution burn brighter and brighter. Pile on the fuel until the forked flames shall lap in wrath around this foul structure of governmental wrong—shall sweep it from basement to turret, and shall sweep it from the face of the earth."[59] Wallace, though he could be irrational in his passions, favored a more grounded, less vitriolic approach, as he remarked to his great-grandchildren in his memoirs: "I never could keep my head if I went much above the earth's surface. (Nor I might say ever since, if I got too far ahead of the convictions of the best people)."[60] Uncle Henry's "middlegroundedness" lent his editorials credibility in the eyes of farm readers seeking to wade through the propaganda of the Agrarian Revolt. Russell Lord reports that H. W. could not stomach the fundamentalist zealotry of the radical farm movements, though he himself sympathized with most of the planks in their platforms. "This may seem strange," Lord reflects, "in a man so religious; but the shouting and twitching manifestations of religion, it must be remembered, he had always held in scorn."[61] It was Henry Wallace's increasing frustration with politics and political ends, his enduring, fond memories of his own college days, and his newfound interest in his son Henry C.'s agricultural education at Iowa State that would cause him to turn, increasingly, to education and to young people, rather than politics, as the source of his legacy.[62]

The appointment of Henry C. Wallace, Uncle's Henry's son, as assistant professor of agriculture at the Iowa State Agricultural College formalized the connection between the Wallaces and higher education, while planting the seed for arguably the most popular farm periodical of all time, *Wallaces' Farmer*. During his years as a junior professor, H. C. Wallace began publishing a practical experimental ag bulletin called *The Farm and Dairy*, which carried in it the seed for *Wallaces' Farmer*. Meanwhile, Uncle Henry, wise but still in many ways rash, neared bankruptcy as the depression of 1893 took hold. Despite owning nearly a thousand acres of land as well as stock in *The Homestead*, he could not pay the notes on his mortgages. These "dark days"[63] would, in hindsight, provide fodder for Uncle Henry's *Letters to the Farm Boy*, though he did not know it then. Wallace's health was failing him, too, in 1894, causing his doctor to prescribe rest from "thinking and talking in public,"[64] a tall order for the verbose Wallace.

Following doctor's orders, H. W. and his wife sailed for Europe in 1894 and, aboard a transatlantic steamer, Wallace's most popular book, *Letters to the Farm Boy*, was conceived when he noted on the passenger list an old schoolmate, a man Wallace insured anonymity in assigning him the allegorized name "Thomas Hardman, Esq."[65] The conversation that followed likely formed the genesis of the dialogic *Letters*, as Hardman, wearied by Uncle Henry's righteousness, reportedly said, "Henry, you've been a fool all your days...a fool on a fool's errand. You have helped wise men into place and power, and they have kicked you; you have given scoundrels your confidence, and they have betrayed you, and slandered and abused you in order to make themselves believe that they owed you nothing. You and I are as far apart as the poles."[66] H. W.'s old schoolmate, Hardman, had indeed been hardened by the world, and served, therefore, as a convenient foil for Henry Wallace's benevolence: Hardman played Mr. Potter to Wallace's George Bailey. A year after this based-on-real-life encounter with his dialogic opposite, Wallace began writing *Letters to the Farm Boy*, a book that Russell Lord described as a book of "parables."[67] More accurately, the book is a mixture of epistle, sermon, personal essay, advice, allegory, morality play, cautionary tale, and autobiography, an innovative, mixed genre that allowed Uncle Henry to be wholly himself and, at the same time, larger-than-life. The characters of *Letters to the Farm Boy*, though, are not merely allegorical constructs. The "Goodmans" are said to represent, in spirit, Wallace's matrilineal line (the Rosses) while the "Brodheads," another emblematic people Wallace names for praise, originate in the maiden name of his daughter-in-law, Henry C. Wallace's wife.[68] Uncle Henry's description of the Brodheads as too proud for "practical politics; that is, office getting and holding," "blunt in speech," and "deficient in holy zeal at revival times, and political zeal during campaigns"[69] makes explicit the equation between the Brodhead and Wallace families. So great was the supposed autobiographical slant of *Letters to the Farm Boy* it led many to mistake art for artist. When Wallace wrote convincingly about conflicts between farm fathers and sons, for example, many assumed he referenced his own domestic affairs. Years later, on the occasion of their father's passing, Henry Wallace's sons were keen to dispel that long-standing rumor, remembering, for the benefit of the readers of *Wallaces' Farmer*, "To him, a quarrel between father and son was a terrible thing, and his mostly widely read book, *Uncle Henry's Letters to the Farm Boy*, was inspired by just such a quarrel which he chanced to witness in a friend's home."[70]

Despite the fact that his health was a wreck, his finances were in shambles, and his editorship of the *Iowa Homestead* was terminated with

three days' notice, Uncle Henry would describe 1895, the year he began writing his letters, as "one of the most memorable in my life."[71] Like George Bailey, Henry Wallace had fallen from grace. Though his readers still adored him, ever the more so for his principled, contrarian stands, he was a horseman without a horse: an editor without a newspaper. Though he had once been, arguably, the single most powerful man in Iowa, he was now a man without any means of sounding the rallying cry. In the Biblical sense, he was in the wilderness, enduring a time he called his "greatest trouble"[72] but also, somehow, his "greatest success." His sons later adopted not only the terminology their father applied to what contemporary poet and mythologist Robert Bly calls the "time of ashes," but wholly embraced the American notion that "times of trouble"[73] reveal true character. Uncle Henry had almost single-handedly put men, including Iowa Governor Larrabee and Congressman Wilson, into power, and yet he could not help himself. He was sixty years old and, as the saying goes, "washed-up." Russell Lord reports that those who knew him at the time, called him a "lion in torment."[74]

Perhaps *Letters to the Farm Boy*, in rekindling Uncle Henry's highly moral, "Fighting Scot" nature, helped reinvigorate him. Certainly, it reminded him of the necessary interdependence of farm families like his own. Resolved to fight, the elder Wallace borrowed money from friends and took his former bosses at the *Iowa Homestead* to court to regain the value of stockholdings refused him at the termination of his editorship.[75] He also sued the owner of the *Homestead*, J. M. Pierce, for libel. And though he proved victorious in both cases, he took no joy in the legal machinations. In his memoirs, he writes that he learned an ultimate lesson about the legal system: "even if you win, you usually lose."[76] Wallace's reminiscences, written to his unborn great-grandchildren, exhibit more sobriety at, than pride in, his legal redemption. He writes, "If you great-grandchildren are interested, you will find the story in all its details in the records preserved by the family. It is a most interesting story, but I don't know whether you will gain much by reading it. I doubt whether anyone gets very far ahead by dwelling upon the mean and discreditable side of human nature."[77]

Meanwhile, Henry Wallace had once more found an organ through which he could reach the nation's rural readers, assuming the editorship of the *Farm and Dairy* paper that his son, Harry C. and Professor C. F. Curtiss, both on the faculty of the Iowa Agricultural College, had purchased. After buying out Curtiss's share, the paper, now called *Wallaces' Farm and Dairy*, asked readers of Wallace's former paper, the *Iowa Homestead*, to send in their names and contact information. The result—thousands of freely supplied subscriber addresses—was a de facto vote

of support for Uncle Henry; he savored the vote of confidence. "I had always great faith in the Iowa farmer and his sense of justice," Uncle Henry wrote his great-grandchildren, "but I confess that my highest expectations had not allowed me to hope that they would so quickly and vigorously rally to my support."[78] With his sons Harry and John, the latter of whom had quit his studies at the Iowa Agricultural College due to worsening eyesight, overseeing the business operations in Ames, Uncle Henry wrote his editorial copy from Des Moines, where the Wallace plant moved in 1896. Though by that time Wallace reported that "it seemed quite clear that we were going to make a success of it," the Wallaces endured two years of personal attacks from Pierce at the *Homestead*, mounting legal bills, and what Wallace alleged was the "stealing of the subscription list" and a "campaign of vituperation and falsehood that was carried on both through the *Homestead* and by word of mouth through some of its advertising solicitors."[79] As the Wallace sons would later write, their motivation in starting *Wallaces' Farmer* was both logical and instinctual. Their account, written in the aftermath of Uncle Henry's passing, reads very much as a tale of redemption:

> When the time of trouble came to him [Wallace], twenty-one years ago, we had not been working together. The father had been engaged in the publishing business with others. One son was in college work, a second had just reached man's estate, and a third was still a boy in high school. *Wallaces' Farmer* was started at that time. The father and the two older sons came together in the effort to repair the family fortunes, to vindicate the family name, and to establish a farm journal which could always be depended upon for certain principles, which soon afterward were expressed in the words that have ever since been carried on our front page, "Good Farming, Clear Thinking, Right Living." It was, therefore, not a case of the sons following into the father's business, but of father and sons coming together to establish a business of their own. And with all of us, the motive was higher than that of mere moneymaking."[80]

The Wallace boys' reminiscence evinces the all-hands-on-deck nature of the enterprise. The spirit of barn-raising pervades. Harry C. resigned his professorship to join the cause. John Wallace, according to Russell Lord, "covered all of Iowa on a bicycle soliciting livestock advertising and subscriptions" while Daniel, the youngest of the Wallace boys, served as "his father's secretary and field man."[81] In an era when women farm writers were few, Uncle Henry's wife, ("Aunt Nancy" as she would come to be called) wrote about domestic life on the farm.

"Good Farming, Clear Thinking, Right Living" served as catchwords for the new magazine, and a distillation of the Wallace family values. The late 1890s brought the Wallace Publishing Company their first, hard-won notices in major midwestern newspapers, as the prolific elder Wallace penned a growing library of farm publications. The 1899 *Milwaukee Journal* lauded Uncle Henry Wallace's short monograph "Trusts and How to Deal with Them" as "careful, conservative, and thoughtful.... free from clap-trap and epithets and...close to the matter."[82] The *Journal* review and others like it pinpoint the exact moment in time when the virtues of Wallace the man and Wallace the writer coalesced to win praise on a regional scale.

Daniel Wallace, the youngest of the Wallace sons, remembers the family's coming together, and the relatively rapid, but hard-won success that followed. He writes:

> The happiest years were along there in the early nineteen-hundreds, with all the lawsuits settled, a paper of our own going, no business or political brawls to upset Father's digestion or trouble the goodness of his spirit, and with money coming in regularly. The agrarian uproar had died with the Populist and Granger breakdown. Now everything was farmers' institutes, Chautauquas, Acres of Diamonds stuff—self-improvement. It was a big time for Gospel Seed Corn Trains, and Co-operative Creamery demonstration tours. Father loved it. He and Harry helped to organize trains and demonstrations....And Mother, with the family pretty well grown up and out from under her foot, was having just as fine a time with her women's club work, advancing culture."[83]

By the time the Golden Age of agriculture arrived, Wallace the Elder was nearing his seventieth birthday and, as happens in so many families, a "chip off the old block" came along in the form of his grandson, Henry Agard Wallace, future founder of Pioneer Hi-Bred International, Secretary of Agriculture, Secretary of Commerce, Vice President, and Progressive Party candidate for President, a man who would be named, ahead of Herbert Hoover, the "most influential Iowan" of all time by the *Des Moines Register*.

Born in 1888, Henry A. Wallace was, by the early 1900s just coming into manhood, a veritable spitting image of his grandfather, Uncle Henry. Russell Lord describes a proud, opinionated Henry A. Wallace who thought the Des Moines suburbs were "putting on airs"[84] and, just as his grandfather before him, "maintained a manner and appearance aggressively rustic, to his mother's dismay." Thus, though the first edition

of *Uncle Henry's Letters to the Farm Boy* had been published by the time Henry A. entered the world in 1888, the letters accrued, as Henry A. grew to manhood, a familial as well as a universal resonance.

Henry A. Wallace, Russell Lord reports, invoked his father's wrath, when as a fourteen-year-old budding botanist, H. A. experimented with potash fertilizer by surrounding the beautiful family cherry tree with an ugly ring of wood ashes. As if he had written a personal letter to Henry A. gently advising him of the error of his ways, Uncle Henry writes in *Letters to the Farm Boy*:

> When your Uncle Henry was a boy, he was very anxious to get over a great deal of work. For instance, he was anxious to be the fastest corn husker and the fastest grain binder in the neighborhood. Unfortunately, he formed the habit of binding sheaves loosely, and failed to acquire the habit of getting all the silk and husks off the corn. The mice had a picnic in the corn that he husked. A loose sheaf when hauled in, or out at threshing time, was instantly recognized as one of "Henry's sheaves." I tried hard to correct this habit in after years, but never succeeded. I could bind tightly enough as long as I kept thinking about it; but the moment I began thinking about something else, and that was about all the time, the sheaf bound itself loose.[85]

As he grew, Henry A.'s politics, even his demeanor, resembled his grandfather's. Just as Uncle Henry had championed conservationist causes as a young newspaper editor, especially the importance of cover crops and grasses, so did Henry A. In his sequel to *Letters to the Farm Boy*, entitled *Letters to the Farm Folk*, Uncle Henry writes with grandfatherly interest in the farmer new to the land. "We watch to see how he will farm," he writes. "We mourn when we see him plow up clover and bluegrass pastures and convert them into cornfields, and there is no lowing of cattle in the empty barns or stables, but robbery year after year of the soil fertility which our old friend has stored."[86] As an older man, Henry A. Wallace would echo his grandfather's concern for the land, quoting Arthur Mason's assertion that the plow was "bleeding to death" midwestern soils. Only a thoroughgoing replacement of "clean-tilled corn culture"[87] with leguminous plants would provide a cure.

Corn, king of Iowa's economy, would be the ticket for Henry A. Wallace. In his first, now legendary teenage field experiments in 1904, he would dispel the widely held belief that the quality and virility of an ear of corn could be predicted by the visible health and girth of its seed. In proving the Shakespearean adage "all that glisters is not gold," Henry A. Wallace would also redeem, at least in a metaphoric sense, his and his grandfather's unconventional nature. As "seeds" of the Wallace line, they were exotic ones. In describing young Henry A. Wallace as "a thin boy whose face was too earnest for his age,"[88] corn expert and Henry A. Wallace mentor P. G. Holden might as well have been speaking of Uncle Henry.

In the very years in which his namesake grandson published his first articles on corn genetics in the family paper, the irrepressible Uncle Henry Wallace, quintessential late bloomer, likewise thrived. In fact, Uncle Henry, now seventy years old, had landed his biggest fish yet in 1908 in his appointment by Theodore Roosevelt to serve on the Country Life Commission, "the great galvanizing event in the history of the rural reform movement" according to Country Life scholar William L. Bowers.[89] The six-man commission amounted to a who's who of American conservation and agriculture, with Uncle Henry as its senior member. In its totality, the Commission consisted of "an impressive group of academicians, journalists, public servants, and practical organizers"[90]; in Henry Wallace, Teddy Roosevelt hand-picked a man who embodied each of these roles. The Commission's whirlwind, fact-finding tour would eventually cover twenty-nine states, thirty public hearings, and the solicitation of more than 100,000 survey responses.[91] None of it phased the elderly Wallace.

In keeping with his advice in *Letters to the Farm Boy*, Uncle Henry and his son Harry, once more a team in raising the precocious next generation, presented Henry A. Wallace with graduated levels of responsibility but withheld the silver spoon. H. A. paid his own way through his senior year at college by working as a correspondent for *Wallaces' Farmer* in addition to ongoing work in the lab.[92] Like his grandfather and his father, Henry A. Wallace would, in keeping with the counsel given farmers' sons and daughters then as now, balance academics with work in the "real world." Though he graduated at the top of his class at Iowa State according to Russell Lord, Henry A. expressed a desire to earn a living at the seed corn business before pursuing graduate work.

As Uncle Henry and son Harry looked on proudly, Henry A. made it safely through each of the difficult stages outlined by his grandfather in *Letters to the Farm Boy*, including courtship, which resulted in H. A.'s betrothal to Ilo Browne in 1914. Uncle Henry followed his grandson's courtship closely, writing to his son Daniel, "Henry has a girl, and it seems to be very serious."[93] Indeed the relationship was serious, quite literally, as Henry A. Wallace is said to have read *Farmers of Forty Centuries* to his wife-to-be when first he went calling, rather than the expected romance novel or romantic poetry.[94] Here again, grandfather

Wallace had divined the farm boy's nature. In "The Farm Boy and His Start in Life" he wrote:

> You will not get the right kind of a start by going in debt for a courting buggy, to spend your evenings in going to dances, circuses, etc., with some good-looking girl, who, if she would speak out, does not value you above one of her hairpins, who eats your caramels and ice cream, thinking, if she thinks about you at all, that you are a silly goose for wasting your substance in that kind of entertainment. She more than half suspects that the buggy is not paid for, she knows you are wearing more stylish clothes than you can afford, and she secretly makes up her mind that while she will have all the fun she can with you, she will say "Yes" to an entirely different sort of fellow.[95]

While his father Harry had courted with fine horses, Henry A., like his grandfather, preferred homelier methods.

In *Letters to the Farm Boy* and its sequel *Letters to the Farm Folk*, Uncle Henry perfected an appealing style. Part sermon, the letters reflected his past as a preacher. Part lecture, they revealed a widely read, college-trained intellectual; direct, they evinced a journalist's love of the straight scoop; folksy and chock-full of horse sense, they belied a man of the soil as well as a man of the people; experienced without being judgmental, the letters revealed a man mellowed by age. The ethos Uncle Henry created in his letters was genuine; in "real life" Wallace insisted on an open-door policy even while at work at *Wallaces' Farmer.* Both his biographer Russell Lord and his sons John, Harry, and Daniel recall that Henry Wallace Sr. routinely dispensed both money and advice when people arrived at his door in desperate straits. "All his days he had been an easy mark for nearly anyone," Lord writes, "however shiftless, dissolute, or abandoned, who came asking alms outright or a loan which, by reason of the nature of the borrower, would never be repaid."[96] When his sons gently questioned his charity, their contrarian father only formalized it, calling his banker and drafting a trust that would pay a modest stipend to himself and his wife in their old age, a small annual amount to his children, and the remainder to particular causes and charities, most having to do with children. Uncle Henry's generosity helped fund the Wallace School in Des Moines, whose students would attend their benefactor's funeral by the dozens, laying roses at his casket. On the first anniversary of his father's death, Harry Wallace, still at the helm of the paper he helped found in the family name, recalled for *Wallaces'* readers how his father had once come to Harry's office, puzzled by the material expectations others had of him. Harry recalls Uncle Henry saying, "I tell them that I am the richest man in the country, that I have all the children I want, all the grandchildren I want, and all the money I want, and I don't care to make it any more; but they don't seem to believe me."[97]

Henry Wallace's persona as Uncle Henry took shape in the early years of *Wallaces' Farmer*, when, finally free from financial woes, H. W.'s interests turned towards what he loved best, counsel-giving. In a tribute to their father published after his death, the Wallace boys remembered:

> When some of the young folks of the *Farmer* office decided to set up homes of their own, he performed the marriage ceremony. When death entered their homes, he conducted the funeral services....It was a delight to him to help in time of trouble. His office long since became a place of refuge to those who sought help and sympathy. And he was the safe repository for the most intimate correspondence from many who never met him face to face, but, from having read his writings, knew he could be trusted.[98]

By his employees, Uncle Henry was as beloved as by his readers. His staff, contributing an unsolicited note to the book-length volume of post-humous tributes published by the Wallace Company, eulogized, "He was familiarly known as 'Uncle Henry,' and such was the term we used when addressing him or speaking of him. Such an appellation as 'the boss' was never dreamed of, and would have been considered almost a sacrilege."[99] The employees go on to remember a man who "drew no social lines" and who "expected you to have honesty, integrity, and a clean mind." "If you had these," his workers remembered, "you were considered by him a social equal, regardless of your financial condition."[100]

With grandchildren growing around him, Henry Wallace had come full circle, back to the abiding humanistic principles and firm belief in youth that had led him initially to the schoolroom and pulpit. As a writer, too, he once again embraced the "expository style" of his preaching days, developing a trademark tone at once plainspoken and philosophically rich.

Uncle Henry's Letters and Lessons

What would become Uncle Henry's calling card, his weekly Sabbath School Lessons in *Wallaces' Farmer*, happened almost by accident, after he had read some poorly prepared Sunday school lessons in a religious magazine. Wallace recalls:

It occurred to me then: Why not supplement these lessons by such a description of the times and characters as would put some human interest in them? I had no thought then of making them a permanent feature of the paper. I wrote one, which sounded pretty good to me, and put a note in the paper that they would be continued for three months, then discontinued if they were not acceptable to our readers. They at once sprang into popularity, and I have continued them ever since.[101]

From that point forward, farmers would frequently include notes in praise of Uncle Henry's lessons when renewing their yearly subscriptions. Featuring Wallace in his 1927 book *Pioneer Agricultural Journalists*, William Edward Ogilvie writes, "not only did his [Wallace's] readers follow him as a leader in aiding the improvement of their business, but also for his gospel. Some farm communities which in the early days could not support a preacher adopted the custom of appointing one of their group to read Mr. Wallace's weekly sermon for the Sunday service."[102]

The stories documenting the popularity of the Wallace's weekly lessons have themselves become legendary. Henry Wallace recalls, in his memoirs, a farmer who came into his office to ask that the lessons continue. His wife, it seems, did not allow him to read anything but the Bible on Sunday. However, the farmer reported, "When she sees me reading the *Wallaces' Farmer*, I am always reading the Sabbath school lesson."[103] Almost apocryphally, the Sunday School Lessons became a Bible for rural folks across the country. The *Des Moines Capital* newspaper, among others, confirms the Wallace family account of the series' popularity, noting the serial's "nation-wide reputation for excellence," while Harvey Ingram, then-editor of the *Des Moines Register and Leader* wrote that Wallace's words were read by nearly half a million people every week.[104] Russell Lord goes further in his definitive biography, gushing, "Never probably in the history of the American farm journalism was there another circulation builder and circulation holder to compare with it."[105] The paper, Lord notes, had to copyright the lessons, to prevent "reprint by rival concerns"[106]

A *Des Moines Capital* account of the era attests to the charm of the Sabbath School Lessons. The *Capital* story quotes the following account by Des Moines resident Paul Jones:

I was down in a little southern Iowa town, in the interests of the Laymen's Missionary Movement....Thinking to brush up a little, I asked the woman proprietor of the hotel for a Bible. She looked embarrassed a moment, and admitted that there was no such book

in the house. But suddenly her face brightened up. "I'll tell you what I'll do," she said, "I'll get you a copy of *Wallaces' Farmer*."[107]

Henry Wallace's love of the earth and the health of its people naturally extended into his wholehearted support of, and expertise in, conservation and environmental stewardship. Here again, a man in his seventies proved to be in the vanguard, holding his own as a conservation pioneer alongside the younger Teddy Roosevelt and his chief forester Gifford Pinchot. As a testament to Uncle Henry's conservation credentials, he was asked to preside over the National Conservation Congress in 1910. Herbert Quick cites the following, prescient passage from *Letters to the Farm Folk* as proof positive of Wallace's environmentalism: "Growing grain for sale off the land starves the soil. I am speaking now for the voiceless land. It will not feed you unless it is fed."[108] While Henry Wallace had been an innovative, conservation-minded man of the soil since his salad days as a clover farmer, the public's acknowledgment of him as a champion of conservation arrived belatedly, corresponding with the nationwide interest in conservation during and immediately following the Teddy Roosevelt administration. Featuring Wallace in an article entitled "A Leader with No Ulterior Motives," Ray Stannard Baker, writing for the *American Magazine* in 1911, quotes the *Des Moines Register and Leader* description of Wallace as a "conservationist who knows something of conserving human resources as well as the resources of forest, stream, and field."[109] By 1935, when James E. Boyle wrote "Our Three Wallaces" for *The American Mercury* magazine, Henry Wallace the conservationist was credited with having "put the word 'conservation' into general circulation and made the whole nation forestry-minded."[110]

Despite his national prominence in all matters of the soil in the last decade of his life, Wallace refused to rest on his laurels. So, while he affirmed farm boys and girls in his letters, he charged them, as he did all young people, with great responsibility, so high were his hopes for American tillers and their families. Addressing the Y.M.C.A. in Des Moines, Clifford Thorne quoted Wallace's grand vision for America's rural people. Wallace, Thorne reports, had once said to him:

My aim is to develop the agriculture of the nation, and especially of the west; to aid in developing a class of farmers mightier than Caesar's legions, more invincible than Cromwell's Ironsides, the stay of the country in war, its balance-wheel in peace, when other classes lose their heads. I wish so to live and work that when I am dead and gone, my name will be remembered by thousands as a man who has left the world better than he found it.[111]

In his later years, Wallace traveled the country as a lecturer and a speaker. When speaking to the next generation, he adopted a tone both challenging and caretaking. During a Farmer's Week student convocation at Cornell University in 1912, he cautioned the mostly Eastern, well-to-do student body against smug complacency. In the address one also hears Uncle Henry's own efforts to keep himself down to earth:

> You will make a great mistake if you imagine that by pure intellect, by pure eloquence, by great ability as a writer, or an organizer, or an executive, you can do anything that is really worthwhile. You may acquire fortune, fame, the power of adjustment over the lives and good fortunes of men; but as soon as your grave is nicely rounded, and you have furnished the text for the penny-liner in the newspaper; before there is time to rear over you a monument of marble, the world will forget you; or, if it remembers you, it will be only to throw stones and flings curses at your monument, if you have been its oppressor.[112]

In this same speech, Henry Wallace, Sr. seems also to be speaking to his grandson, future Vice President and world-renowned geneticist Henry A. Wallace, at a time when the scientific community, and increasingly the general public, was consumed with heredity. As the scion of a multigenerational American family that was, in its era, arguably second only to the Roosevelts in multigenerational political and social prominence, Uncle Henry Wallace dismissed the notion of genetic superiority, especially when applied to his own family. And yet this line of inquiry, which so displeased the progenitor, would grow more popular as Henry C. Wallace would be named Secretary of Agriculture in 1921 and his son, Henry A., would go on to serve in that same capacity in 1933. As if cautioning his own descendants, particularly his top-of-the-class grandson Henry A., Uncle Henry further took the Cornell student body to task in his convocation speech:

> It will be fearful to assume that an additional sheepskin, even with cum laude or summa cum laude, will insure your success in the twentieth century. You have this great advantage in entering upon your life work in the twentieth century: that if by making the best of your heredity and environment you become really big, your power will be vastly multiplied by the machinery of the organization, just as the power of the farmer and the mechanic has been multiplied for the past half-century. For this is a century of big things—big trusts, big banks, big railroads, big newspapers, big universities, a vast and complicated network of organizations; and the man who is big enough to go to

the head and hold his position has a power for good, and also, alas, for evil, hitherto unparalleled in the history of the human race."[113]

Resisting notions of genetic privilege, H. W. opened his speech by sending a shot across the bow of the fortunate sons and daughters of Cornell University: "You will risk failure if you take to yourself credit for being born of good old New England stock, or of Dutch, or even Irish or Scotch-Irish. They were really good people, and did their work well, without your help, and you are entitled to no credit."[114]

Uncle Henry's "Stud Book"

The public's interest in determining what rare fire burned within the "Wallaces of Iowa" was inevitable. The family's genetic inheritance, after all, had shown none of the diminishment and degeneration of the "good pioneer stock" that many felt characterized American youth in the run-up to the Roaring Twenties. Furthermore, the success of three generations had been not only literary and scientific, but also political and financial, as *Wallaces' Farmer* had become sufficiently lucrative to earn the moniker "Wallaces' Gold Mine." John P. Wallace and the youngest of Uncle Henry's boys, Daniel, likewise enjoyed national reputations as agricultural editors. In his book, *Pioneering Agricultural Journalists*, writer William Edward Ogilvie describes Daniel A., the editor of the St. Paul- and Minneapolis-based serial *The Farmer*, as the "most widely known farm paper director in the northwest."[115]

   Even Russell Lord, biographer and a friend to the family well aware of the elder Wallace's aversion to aristocratic claims, could not resist referring to genetics in his opening chapter of *The Wallaces of Iowa*. Variously calling the three generations of Wallaces "Country Gentlemen,"[116] "Gentlemen of the Midland "[117] and "Midland aristocrats,"[118] Lord traces "the closely linked succession of grandsire, first-born son, and first-born grandson." "The lives of these three men exhibit an extraordinary genetic continuance," Lord declares "of which they were, or are, conscious and proud. They grew close together and grew fast once their clan, transplanted from Pennsylvania, took root in Iowa."[119] Writing in the mid-1940s, Lord notes that the "present Henry Wallace," by which he meant Henry A., "can tell you not only the full name and life span of remote great-aunts and great-uncles, but also how good their teeth were, when in the course of their lives their hair turned gray or fell out, whether they liked to read, and whether they could carry a tune."[120] For his part, Uncle Henry was, as a farmer and a farm writer, naturally intrigued by

such stuff, though he often good-naturedly mocked his wife about her family's assiduous keeping of its own "Stud Book."[121] When, in 1943, genealogical writer and editor Conklin Mann claimed to have proof that Henry A. Wallace, the "second ranking American executive" was related to Winston Churchill and, further, that H. A. was 53 percent Scotch-Irish and 37 percent New Englander, Henry A. replied diplomatically but disinterestedly in the press in a manner that would have made his grandfather proud. "The nostalgic glow and sense of romantic association with great events and gentle persons of the past which genealogical research encourages" Wallace said, smiling according to Russell Lord, ". . . as a geneticist, I know it's bunk."[122]

Family modesty aside, part of what makes the story of the Wallace family so enduring, and what gives further historical and circumstantial weight to Uncle Henry's *Letters to the Farm Boy,* is precisely family ancestry. After all, Uncle Henry's open letters revealed to wider audiences the very same kernels of wisdom he shared with his children and grandchildren. Though exceeding the American interest in the Wallace family, the public's preoccupation with the Kennedy and Bush dynasties offers something of a contemporary analog, as does, in Britain, the tabloid coverage of the royal family. In any case, by the early 1930s, Uncle Henry had been gone well over a decade and yet the genetic theory of "grandfatherism,"[123] holding that genetic inheritance often "skips a generation," caused an unusually intense interest in the achievements of the grandson, Henry A. Wallace. In essence, the success of the Wallaces begged the question—very much in the public mind in the aftermath of the Scopes Trial: Was it nature or nurture that accounted for the affinities between, in this case, grandfather and grandson? Or was it the nurture of Uncle Henry, the time spent and nuggets of wisdom shared, that fueled Henry A.'s meteoric rise?

Noting the "mental and spiritual resemblance"[124] between Henry A. and Uncle Henry in an interview given in 1942 to *Fortune* magazine, Henry A. Wallace acknowledged that he was much closer to his grandfather than to his father.[125] As a child, H. A. spent countless hours reading books of sociology and philosophy to his grandfather, whose home was just a mile away from the *Wallaces' Farmer* offices. Lord writes, "The grandfather, like the grandson, was interested in nutrition or dietetics and experimented on his own person. His mental and emotional or religious life displayed a constant growth beyond confining dogma. He examined platforms and creeds, lay and religious, and spoke his mind regardless."[126] The comparison between the two Wallaces, explicit and implicit, has by now become commonplace. Writing for newfarm.org, Laura Sayre notes Henry A.'s difficulty feeling at home in Washington

D.C, citing his "strange combination of intellect, hayseed, and ascetic"—close variants of the adjectives used to describe his grandfather during his lifetime. "By all accounts," Sayre adds, "Wallace was a pensive, modest, frugal, deeply spiritual person....he ranged restlessly among different faiths and philosophies, including Transcendentalism, Theosophy, and Liberal Catholicism."[127] This very same openness to spiritual and moral inquiry characterizes the letters and life of his grandfather.

The story of the relationship between the three generations of Wallaces proves as instructive as it is moving. Seemingly immune to jealousy or unhealthy competition, each generation, as it achieved national prominence, honored its predecessor. When Harry Wallace was asked to serve as Secretary of Agriculture, he addressed head-on his father's aversion to public office, writing in a 1921 *Wallaces' Farmer* farewell, "It has long been a tradition of our family not to seek political office....In becoming the head of the United States Department of Agriculture, I do not feel that I am breaking his tradition. I did not seek the office directly or indirectly, as a large number of my friends who have wanted to help me will testify." In his outgoing message, Harry passed the torch to his brother, John P., and his still-young son Henry A., of whom he wrote, "He has been in the editorial work for ten years, and has been doing more of it than is generally realized. That he is fully equal to this larger responsibility our readers will discover for themselves."[128] Again, the importance of fathers passing on real responsibility to sons, in the manner recommended by Uncle Henry in his *Letters to the Farm Boy*, is upheld. Demonstrating almost perfect symmetry, in early 1933 Henry A. Wallace would write his own farewell under the heading "Odds and Ends," where he would remember his father's—Harry C.'s—work in Washington D.C., accept weightily the offer to serve as Secretary of Agriculture in the cabinet of F. D. R, and conclude by telling his readers that he would, in the family tradition, try to do his "part" in Washington. "No doubt," he added, "I shall make many mistakes, but I hope it can always be said that I have done the best I knew."[129]

Uncle Henry's Victory Lap

In his last few years, Uncle Henry Wallace delighted in writing his "lessons" to the readers of *Wallaces' Farmer*. He published his second volume of letters, *Letters to the Farm Folk* in 1915, one year before his death. He began dictating his memoirs, *Uncle Henry's Own Story*, to his great-grandchildren, yet unborn, in the form of autobiographical letters, volume one of which would be published by the Wallace Publishing Company in 1917 with the others to follow in 1918 and 1919, respectively. Behind

the scenes, he stayed politically active, arguing against United States involvement in World War I. Having had the ear of Presidents since Teddy Roosevelt, he arranged to talk with President Wilson in the Blue Room of the White House while passing through Washington D.C. In his closing letter to great-grandchildren once only imagined but now made flesh and blood with the birth of the next generation, Henry Browne Wallace, Uncle Henry describes his meeting of the minds with the President. "I told him that this war must end some time," Uncle Henry writes. In his letter dated November 3, 1915, he continues:

> He [Wilson] was the one man who could attract the attention of the world; that probably the time would come when he would be able to suggest as the basis of lasting peace the freedom of the seas and their policing by an international fleet, so that for all time to come, the nations of the world, wherever they might be located, could freely trade with each other without fear of molestation. I said to him that this was only a vision of mine, a dream; that I made it to him merely as a suggestion, saying that it would not be practical until every woman's heart in the warring nations was broken, until the nations themselves were bankrupt. I said to him that when that time came, surely the common people of these nations would not permit themselves to be crushed under an added burden of taxation, if the ends for which the navies were built and maintained could be subserved without it; and that if the people once came to clearly see this, they would overturn the government that insisted on breeding men for the shambles, to carry out ambitions of their leaders.[130]

The message, so apropos to America's twenty-first-century wars, is vintage Uncle Henry, revealing a man capable of offering even the President of the United States a needed lesson.

Henry Wallace died not long after dictating the above reminiscence. Fittingly, he passed on Washington's birthday, February 22, 1916, while waiting to preach as a guest minister at a church in Des Moines. The auspicious date and location of his death only furthered the Wallace legend in the grief-filled days that followed, when each of his many constituencies made a positive example of Wallace, each according to their cause. Clergy praised his religious life and contributions to the Church, while forgiving him his progressive views and secular profession. Schoolchildren praised him as a generous and loving benefactor; editors praised him as an able colleague; politicians and leading public figures of the day, including Liberty Hyde Bailey, Tama Jim Wilson, and Teddy Roosevelt, wrote in with tributes. F. W. Beckman, a fellow journalist, admired how Uncle Henry never "preached down to country folks"[131]; Clifford Thorne, addressing mourners at the Y.M.C.A., quoted Uncle Henry's thoroughly egalitarian precept, "Every man, whether in private or public life, should endeavor so far as possible to give equal opportunity to every citizen, and to secure and enforce a square deal between man and man."[132] Liberty Hyde Bailey wrote a short but compelling eulogy describing Uncle Henry's invaluable work on the Country Life Commission Bailey chaired. Bailey writes of Wallace:

> The oldest man on the commission, nevertheless he was tireless and always ready for new hearings and to meet new people and consider new problems. He had a certain homey way of interesting the people that was a constant entertainment and inspiration to us—approaching them with anecdote, parable, inquiries about their crops and affairs—always concrete and direct, without losing vision. All day, he would work, travel continuously, and preach on Sunday—always good-natured, entertaining, shrewd and wise.
> He could express disapproval or disagreement in the most unmistakable way, and yet without giving offense. Novel questions he took up with the eagerness of a boy. This has always been a wonder to me—this facility of a man at his age to readjust himself completely and helpfully to large public questions to which perhaps he had not theretofore given much attention.[133]

One of the larger questions to which Henry Wallace was "readjusting himself" in the last year of his life concerned service on the board of trustees at the University of Cairo. In part his interests were missionary, but his vision transcended, as always, religious dogma. In his last reminiscence, he explains his wish to, through his support of the University, help Egyptians become "engineers, scientists, and teachers."[134] Uncle Henry, perhaps sheepish at the far-flung nature of this, his latest educative and religious endeavor, writes to his great-grandchildren, "You may wonder what I have to do with an enterprise of this kind. It may seem at first thought foolish."[135] And in a passage strikingly relevant to the current cultural moment, Wallace articulates his interest in Islam and its importance to the rest of the world. "Cairo," he reminds his would-be detractors, "is the intellectual center of Islam, of the Mohammedan world, and it numbers two hundred million people."[136] In an era when many American farmers took pride in their stubborn isolationism, Henry Wallace continued to travel, to engage.

As he would have wished, Uncle Henry's lessons and letters did survive him, continuing to run in *Wallaces' Farmer* by popular demand

until 1919. His three-volume memoir, *Uncle Henry's Own Story*, would likewise be released after his death in 1917. And, as Henry A. Wallace moved into successively higher offices in Washington D.C, eventually occupying the desk of Vice President, he would see to it that a picture of his grandfather hung always in sight.

Editor's Note

The following letters from Uncle Henry Wallace have been modestly reordered for greater organization and clarity. Toward that end, the sub-headings Advice for the Farm Hobbledehoy, Cautionary Tales for the Farm Hobbledehoy, Advice for the Farm Parent, Advice for the Farm Couple, and Advice for the Farm Grandparent have been introduced to better signpost intended audiences.

In *Uncle Henry's Letters to the Farm Boy*, chapters appeared in the following order: "The Farm Boy and His Father," The Farm Boy and His Mother," "The Farm Boy and his Temper," The Farm Boy and his Chum," The Farm Boy and His Reading, "The Farm Boy and His Future Business," "The Farm Boy and His Fun," "The Farm Boy and his Education," "The Farm Boy and His Start in Life," "The Farm Boy and His Habits," "The Farm Boy from Home," "About the Hardup Family," "About the Richman Family," "The Hardman Family," "Commercial Morality," "The Brodhead Family," "Types of Common People," and "The Good Man." With the exception of "The Farm Boy Hobbledehoy— Neither Man nor Boy," which is a letter and letter heading grafted from *Letters to the Farm Folk*, the individual letter titles in this volume's table of contents correspond with the original.

In *Letters to the Farm Folk*, chapters appeared in the following order: "The Scrap Heap for Boys," "Girls for the Scrap Heap," "The Hobbledehoy—Neither Man Nor Boy," "The Mother on the Farm," "The Home Life of Farm Folks," "The Social Life of the Farm Folk," "Improving the Social Life in the Country," "Work on the Farm," "Getting the Boy Started Right," "Friction in the Farm Home," "Proper Brain Food for Farm Folks," "The Health of the Farm Folk," "Getting Well and Keeping Well," "Farm Folks and Their Neighbors," "Religion for Farm Folks," "Farm Folks Who Have Failed," "Growing Old on the Farm," and "Rounding out Life on the Farm." With the exception of "The Scrap

Heap for Boys," which is deployed in Part I as a preface for a thematic grouping of cautionary tales for the farm boy, individual letter titles in this volume's table of contents correspond with those of the original.

Of the Part III selections appearing from *Uncle Henry's Own Story of His Life, Personal Reminiscences*, all but one, volume three's "Corn Growing and the First Corn Train" come from the first volume in the series, wherein Uncle Henry writes with his great-grandchildren foremost in mind. Volumes two and three, with a few notable exceptions, lose some of their epistolary immediacy as they evolve into traditional autobiography and memoir.

Of the more than 225 pages of letter tributes that arrived in support of Uncle Henry after his death on February 22, 1916, Part IV, in keeping with the theme of this volume, features only those letters and compositions written by Uncle Henry's favorite audience: children—in this case, both his own children and the pupils who attended the school that bore his name.

In sum, the table of contents for *Uncle Henry Wallace: Letters to Farm Families* reflects editorial selections towards a "best of" compendium, though a majority of the original letters are here included. Letters were selected for their relevance to today's rural families. As a guiding editorial principle, the original language of the letters has been painstakingly preserved; in the few instances where substantive alterations were needed, they are accompanied by footnotes. Spelling, punctuation, and mechanics have been modernized only where necessary for readability. Few changes proved requisite.

Full bibliographic citation for each of the books excerpted herein is available in the acknowledgments to this volume and in the notes section.

Part One

Letters to the Farm Boy

Sincerely your Friend

Henry Wallace

Prefaces to Uncle Henry's Letters to the Farm Boy

Uncle Henry's Preface to the First Edition:

> Twenty years hence the farm boy of today will mainly control the business of the state and nation, as it is now controlled by the farm boy of twenty-five years ago. To aid in starting this farm boy on the right track and make his pathway plainer and easier, is the object of this publication in its present form. I have known how the farm boy feels, for I have experienced his isolation, his hopes, his ambitions, his lack of experience and knowledge of the world, and hence I know his need of a kindly, sympathetic friend outside of the family, who will suggest rather than advise, guide rather than lead, who would rather commend than censure, and who is a boy in feeling though a man in years and experience.
>
> Nothing was further from the writer's thought at the beginning of these letters than to write a book. The first was merely an effort to make matters smoother between a father and his son. The rest followed; I scarcely know how. This book wrote itself; like Topsy, "it growed." The marked favor with which the letters as first published in *Wallaces' Farmer* have been received, and the desire expressed on every hand to have them in permanent form, leads me to hope that it will do its part in fitting the farm boy for his high destiny. The farm boy with his robust health, his independent spirit, his training in the primary virtues of industry, economy, and uprightness, and his opportunities for clear thinking, may be the ruling power in this nation if he is rightly guided. To do his part in guiding the farm boy aright is the desire and ambition of Henry Wallace.

Uncle Henry's Preface to the Third Edition:

> While the following letters were written especially for the farm boy, the record of the sale of the two previous editions proves most conclusively that the farm boy's father is as much interested in them as the farm boy himself. Nor should this be a matter of surprise, for in reading them he lives over again what is ordinarily the happiest portion of his life.
>
> It was at first a great surprise to me that the city boy and his father should read these letters as eagerly as do their country cousins. The obvious explanation is that the city boy sees in them glimpses of a life among the green fields and flowing brooks—filled with

convenient swimming holes—for which his soul naturally longs; while to the city boy's father they reveal glimpses of past experience, or of what he has long regarded as an ideal human life.

Advice for the Farm Hobbledehoy

My Dear Boy:

It has occurred to me that matters might not be going just exactly right between you and your father, and that a word from one who has been both farm boy and father might do good to both of you. I do not think for a moment that there is anything seriously wrong, only that neither of you are as happy in your relations with each other as you ought to be and can be. I take it for granted that you love and respect your father—not quite in the same way that you love your mother, because the affection that you bear to the one is distinctly different from that which you bear to the other, and must be in the very nature of things. I take it that you have a good father who loves you dearly and who above all things else desires that you be a strong, true, brave, noble man, who will bear his name with honor when he is lying in the grave. I know he thinks more of you than he does of the farm and all that is on it, saving always your mother and your brothers and sisters. I take it that you are a good young man, and there is no reason why you and your father should not be as happy together as people can be in this world. If you are not, it is likely that both of you are somewhat to blame, and I will venture a guess as to why you are not as happy as I would like to see you.

You, perhaps, think your father is needlessly exacting in some things. He wants that stable cleaned out promptly and thoroughly, and wants the pigs fed just so every time, whether it is wet or dry, or a good day to go fishing or a bad one. He wants the cows milked clean, does not want any loud talking while milking, and he wants the milk cared for just so, and if you fail in any of those things he does not like it, and you do not see why he should be so particular. Now, I will tell you why. Your father was probably a little bit careless himself when a boy; he sees the mistake; he knows how difficult it was for him to get over this habit, and he does not want you to have the same kind of trouble.

You do not see why he disapproves of your going out with a lot of other boys whom you regard as good fellows, but who have some bad habits, such, for example, as using profane language or indulging in obscene talk. Now, I will tell you why he does not want you to go with those boys. He possibly went more or less with that class of boys himself and knows from experience that they are not the kind of boys with whom you ought to associate.

He objects to your going out at night unless it be to some literary event, or to make a social visit to your neighbor. Now, he is perfectly right about this because he has had experience and you have not. You do not see why he insists on your going to church every Sabbath and to Sabbath-school, even if you are tired and sleepy and would like a good, long day's rest. Again I tell you why. He felt when he was a boy just as you do, but years have taught him the necessity of acquiring steady and regular habits of industry, morality, and religion. Your father has lived a long time, has had lots of experience, and knows a great deal that books can not teach, and he would like above all things else to be able to impart that experience to you, which he knows that he can not impart except by insisting on your acquiring it by the doing of it. That is the only way that anything worth learning can be learned. In all these things your father is exactly right.

You perhaps feel that he ought to give you a chance to earn something for yourself—that there ought to be something on the farm which is your very own, or, as your sister might say, your "ownest own." Well, I think so, too. I think you are entirely right in this, and if I were in your place I would, some day after supper when he was not troubled in any way, talk the matter over with him in a manly, open way. Nothing pleases a father so much as to see his boy develop manliness. I would talk to him about this, but I would make a square bargain that if you are to have a pig, or a calf, or a colt on the terms agreed on, it is to be your hog, or your steer, or your horse when it is disposed of, and you are to be the sole judge, after asking his advice, as to how you are to use that money.

You think your father should not bind you down so closely as to the plan you are to take in doing certain things about the farm. You want to exercise your own judgment, and have, so to speak, a little leeway. You are willing to do the things he wants you to do, but you would like to do a little planning and thinking for yourself as to the way of doing them. Here you may be right and again you may be wrong, but I think he had better say to you, "My son, there are certain results that I want accomplished: I think you had better do this way, but if you see a better way, try your hand." You will probably find that his way is the right way after all, but it will do no harm to find that out by experience.

You may think that your father is a little of an old fogy in some matters connected with farming. There is a possibility that his long years of experience enable him to see through the fallacy of theories that you may not be able to do as yet. Therefore, I would advise before condemning his ideas to study them quite thoroughly and weigh carefully what you may see on the other side. He may not be able to give you as good reasons as you may see on the other side on paper, but I suspect that he has the common sense of it pretty firmly fixed under his gray hairs, and may not have the patience to sit down and argue the thing out with you.

I would like for you to have profound respect for your father's views on all questions. They may be wrong—no doubt many of them are—but you should remember that "knowledge comes but wisdom lingers." It may be that you know a good deal more than your father. If so, it is because you take after your mother, but whether you really know it must be clearly established by actual results, and not assumed.

In order to have a proper respect for your father, you must not call him "dad," or "pap," or "pa," or "the old gent," or "the governor," as I have heard a good many English boys call their fathers. There is but one name that he is entitled to, and he is entitled to that every time; and that name is "father," never "the old gentleman." The very act of calling him father will make you respect him and respect yourself, and smooth out any little trouble that there may be between you. It is essential to your growth and future happiness that you and your father have the most perfect understanding with each other. By and by he will come to trust you implicitly. First, he will be to you a sort of older brother, and as the years go on he will learn to depend on you, to lean on you, so to speak, and by and by will be disposed, when he begins to lean heavily on his staff, to pay as much deference to your opinion as you did to his when you were a little boy. You thought then that father knew it all. He will think after a while that you know it all, and that whatever you do is about right because you do it.

I write this to you because I have known boys who took a different course from that which I advise you to take, and who have blighted their own lives and their fathers' lives, and broken their mothers' hearts, and I do not want you do to either.

Affectionately,
    Uncle Henry

My Dear Boy:

You will, I am sure, pardon me if I venture to write to you on some matters that are in a manner sacred, and I do, solely because I believe I can do you some good for which you may thank me ever afterwards. Your Uncle Henry is now over sixty years old, and can, therefore, talk to you as he would not have dared to do twenty years ago. He has all his life had much to do with boys, has boys of his own, and thinks that a bright boy, clean in life, in word and thought, is every whit as noble and admirable a character as a bright, pure-minded, beautiful girl. He has all his life noticed that a boy of this class has almost invariably a good mother, and more than that, that he is a good mother's boy as long as his mother lives. You have no doubt read the account of the inauguration of President McKinley and you know that his tenderness toward his aged mother lifted him in your opinion much higher on that occasion that anything that he said in his inaugural address. You have often heard it said of some bad man, "There must be something good about him after all or he would not be so kind to his mother." I can assure you right now that your whole afterlife will depend very much on the way you treat your mother. In all past ages men have noted this fact. "Honor thy father and thy mothers," said Paul, "which is the first command-ment with promise." That promise was, "Long life and life and prosperity to such as kept this commandment," and noting the fact that disobedient boys come to a bad end, an inspired writer said: This is a simple statement of principle that there is a very close and intimate relation between a boy's success in life and the filial affection which he shows to his father and especially his mother. It is in the home and in early childhood that we acquire those quali-ties that make us truly successful. We learn to love by first loving our mother, we learn respect and reverence from our father, and we learn to respect the rights of others from our brothers and sisters. From my heart I pity the boy who is either motherless or fatherless, and scarcely less do I pity the only son or daughter. They are all necessarily dwarfed specimens of humanity. Note the little apple in the heart of the blossom. The blossom is the home in which the fruit is enfolded until it is fit to endure the sunshine and the storm. If the blossom is injured the apple never amounts to anything. Ever after it grows to maturity and the blossom has long since fallen away, it still leaves its mark on the apple. So it is in your home life. "Poor boy," we often say, "his father died when he was a baby," or

"He had no mother," or "He was an only child," when we wish to excuse weakness for which there is no other palliation.

You are entitled, my dear boy, to all the good a good mother can do you, but you can never realize this until you are good to your mother, and the one proof of this love will be in the seeing that the good is always done, that she has as little drudgery as possible to do about the farm, and that her mind is ever free from care.

Now, let me tell you of some things that you will be tempted to do. Some boy will wish you to join him in something you know your mother would not approve. He will sometimes sneer at you and call you "mother's boy," and say you are "tied to your mother's apron strings." I would not advise you to knock that boy down, because the sneer is directed at you and you can afford to let it pass, but if he says anything against your mother you have my permission to slap his mouth. Do not let any boy of your age or size say a word disrespectful of your mother. Let her religious convictions, her ideas of duty and propriety, her faults even, be too sacred to be found fault with by mortal man.

You are likely, as you approach manhood, to put too little store in your mother's judgment. When a boy gets to be from sixteen to nineteen or twenty he is apt to speak lightly of his mother's influence. She may not be as good a scholar as you are, may not know half so many things, but your Uncle Henry would take her judgment offhand in preference to yours in all matters that affect character or life. When you get to know women better than you now do, you will find they have a very queer way of guessing at the rights of things and guessing right nearly every time. A man reasons, a woman divines; a man thinks things out, a woman feels them out. Your mother is not infallible nor yet perfect, but she is so nearly certain to be right about matters that affect your character and your life that you can not afford to treat her intuitions lightly. If you do, you will make a mistake.

When you become a man you will have a wife of your own, or ought to. You won't own it to me, but I you suppose you are thinking once in a while about that time. She may be a little jealous that any woman should share your affections; possibly she may not be able to help it, but let me say to you that she knows more about other women than you do or ever will. If you are as good to your mother as you ought to be, she will in the proper time take your girl into her heart and life as a daughter indeed.

You will be all the happier if you make your mother your confidant in your love affairs. Coincidentally I may say to you she is ordinarily about the only one of the family you can advise freely on that subject. Your brothers and sisters might laugh at you, and you do not like to talk to your father about it, but I assure you that you can have no better adviser in a matter which may be happiness or misery unspeakable to you than your own mother. There is something very beautiful and touching in the affection of a mother toward a good boy when her hair is white and her step tottering. His hair, too, may be gray, but to her he is a boy still, repaying in tenderness and kindness and helpfulness that quenchless love which she lavished upon him from childhood to manhood. By kindness and tenderness, by making her your confidant now, you can make your mother the happiest of women and, at the same time, do much to make your own life a success. One little act of kindness shown her each day will do it.

Affectionately,
    Your Uncle Henry

My Dear Boy:

I have not sized you up as a good-goody boy such as too often figure in Sunday school books. Such boys are too often like the apples that ripen too early, indicating that the tree is on the decline. I do not think there is much danger of your dying early on account of being too good for this world. I have seen you get mad and fight and sometimes hear you say words not found in the dictionary. I do not approve of these things; neither will you when you are older; and yet I have more hope of a boy built in that way than of one who is goody-good, and a great deal more than of one who prides himself on his cunning and deceit, or who delights in doing little, underhanded, mean things, such as telling tales out of school, and meanwhile playing the role of a saint. But, my boy, if you are to make your mark in this world you will have to learn to curb that temper, a thing which you can never do until you learn to curb that tongue.

I know how many things there are about the farm that make a boy mad and let his tongue loose at both ends. There is, for instance, the experienced brood sow that will go the wrong way when you drive her, that will not lead worth a cent, and goes about where she pleases. Her pigs seem to lie awake nights thinking how to get

into the garden or potato patch, and when you discover the little rascals, they clear out with an air that seems to say: "Didn't we come it over our bubby?" Then there is the cow that opens the gate as if her horns were hands, and that other cow that kicks on the slightest excuse and generally manages to get one foot in the bucket when it is half-full. Then there is the wise old brood mare that will come at your call in the pasture and take the corn out of your hand, but if you reach for her foretop, will show you her heels, and let you feel them, too, unless you are lively. I do not wonder that you get angry and are tempted to take a club to the sow, beat the cow with the milk stool and whip the old mare—when you get a chance. I have felt just that way many a time. Then, there is the balky horse that looks around over his shoulder when he comes to a soft place in the road or to a little hill, and stops and stays stopped—a regular quitter that will neither be coaxed nor forced to budge an inch, and seems to enjoy your anger. You may be pushing the mower in hay harvest and about every rod or two the sickle runs into a gopher hill, and you have to stop and back, and clean it off, and are scarcely back in your seat until you run into another hill, and this time you say, "Confound the gophers," or perhaps worse. Or, you may be in a hurry to get off a load of hay and get in another before the rain and the horse fork takes a tantrum and drops the forkful too soon or twists around and will not drop it at all, and your father loses his temper and comes tearing in to know what keeps you so long at the barn.

Oh, the farm is a grand place to try a boy's temper. It is almost sure to rain at the very time when you are promised a day's fishing, and the best horse on the place goes lame when you expect to take your best girl to the Fourth of July. At least that is the way it used to be. Nevertheless, my boy, you will have to get the better of your temper or your life will be somewhat of a failure, and the less you curb it the more of a failure your life will be.

For a boy or a girl to get mad and fly off the handle is somewhat excusable; for a man, never—well, hardly ever. "Be ye angry and sin not," said an inspired apostle who, himself once got angry and called the chief justice a "whited wall," which means simply a first-class scoundrel, so I presume there is an anger that is altogether justifiable; at least I hope so, but "let not the sun go down upon your wrath." The boy is not supposed to have gained control of himself; the man is. I know men, and a good many of them, who are very strong in many ways, have nearly every other element of great success but this one of self-control, and they sometimes make stark fools of themselves and lose the respect of their best friends because they fly in a passion on very slight pretexts; or, what is even worse, sulk and pout and then go home at night and kick the dog, scold their wives, if they dare, and their children go off to the barn or to bed for fear of their father's anger. This is what may happen to you when you become a man unless you get control of your temper and your tongue while you can.

Now let me whisper a secret: That cow and the brood mare would not be half so "ornery" if somebody had not been in the habit of losing his temper. That balky horse would never have learned to balk if his first owner had had good horse sense and controlled his temper. I do not say that any measure of self-control on the part of the owner will take the contrariness out of a hog, but it will take away the opportunity of showing it.

You may ask me how to control that temper. Let me confess to you that it is not an easy matter, but where there is a will there is a way and a wise father and mother will help you in this good work. Your Uncle Henry had a furious temper when a boy. He got mad when he was turned down at school, and flew off the handle about something or other nearly every day. He remembers very distinctly his first lesson in curbing his temper. His father put a J. I. C.[137] bit on him one morning very neatly. He did not like something at the breakfast table and, on being reproved, flew off as usual. The first thing he knew he got a dash of very cold water in the face, and then another and another. The shock enabled him to get control of his nerves. He then and there found that he could control his temper if he but tried. It is all nonsense to say that a boy can not control his temper. Did I not see you the other day in a passion when working on the road? The other boys laughed at you and you looked around and saw your best girl coming in a buggy and looking as sweet and cool as a rose after a shower, and in a second you were all smiles and touching your hat to her and felt a little ashamed of yourself all that day. No matter how angry you are when you can hold your tongue when a stranger for whom you have the greatest respect is present. If you can do it with this outside help, you can, if you try, do without it.

Bear this in mind, that sinful anger is never a mark of strength or of manliness, but always of weakness. It is a sign of immaturity—vealishness, if you wish to call it such. It never contributes to happiness and always makes a sensible man feel cheap and mean when he comes to himself. I am free to say that I have never been angry without good cause, and let my tongue loose without after-

wards loathing and despising myself. One cannot afford to lose his self-respect and hence to maintain it is obliged to secure self-control.

One can sometimes do by indirection what he can not do directly. When a farm boy I once caught a preacher whistling on Sunday morning, which I was taught was a very great sin, and ventured to ask him why he did it. He colored up and said, "I'll tell you, my boy, how I got into the habit. I had a fearfully bad temper when I was a young man and resolved to whistle whenever I found my temper rising. I then got into the habit of whistling whenever I was thinking seriously about anything, and I was just now thinking over my sermon, and it whistled itself." Another once told me he got control by counting three before he let his tongue loose on the other fellow. Mind this, if you can keep your tongue between your teeth, you will have little trouble with your temper.

I don't say a man should never become angry. Far less do I say he should not show resentment. There are some things on the farm and a thousand times as many off it calculated to make a true man's blood boil and fill him with righteous indignation, and he ought by all means to show it. Neither man nor boy has any right to stand insult or endure wrong without showing resentment, and that in a most pointed way. A man or a boy who allows another to wrong him or insult him without resenting it lacks something essential in the proper makeup of a true man, and actually becomes an accessory to wrongdoing. I think the recording angel conveniently forgets to report the boy who knocks down the bully who by brute force terrorizes weaker boys. If I had the reporting to do I would look the other way, and if I happened to see it would report a credit mark instead. The public sentiment that justifies a brother in putting a bullet hole through the man who ruined his sister is not wholly evil. It teaches brutes to control their passions and fools to hold their tongues. Resentment however, to be effective, must be with perfect self-control. If you have to show your teeth, do it deliberately, and show just enough and no more. If you bite, do it with perfect coolness and good purpose. You can never do this unless you control your temper. The boy to be feared is the boy who can knock you down and smile, whose eyes are ablaze with fire and yet under perfect control. It is this that marks the really strong man that I wish you to be.

You will be surprised to find how much the habits of the stock on the farm will improve when you get control of yourself. On many farms the livestock unconsciously tell the observant man just what kind of a temper the owner, or some of the boys, or perhaps the hired man, has. Remember what Bobbie Burns said: "Know, prudent, cautious self-control is wisdom's root." So thinks your Uncle Henry.

My Dear Boy:

If you are to become the good and true man that your father and mother hope you will be, it is very important that you choose the right kind of a friend. Tell me the kind of a chum a boy has, and I will tell you what sort of a boy he is and what type of a man he is likely to become.

I sometimes think that it is essential to the right development of a boy that he should have, first, a dog; second, a chum; and third, and last, his best girl. It is a little too soon to talk about "the last and best yet," but if you have fallen in love with the right kind of a dog and selected the right kind of chum, you will not go far wrong on the best girl; and if you do not find her, she will happen in on your path by accident or providence, when the right time comes. Your father knows all about that, I am sure. I like the boy that likes a good dog, a dog that is bright, honest, and industrious—that looks you square in the eye without flinching, and will fight for you when it is time to fight, and there is something wrong with the boy that likes a downright mean, cowardly dog—a dog with a bad conscience.

After the dog, but generally along with it, comes the chum, and he is the making or marring of more boys than parents think.

I like a boy who has one particular friend about his own age, a friend or chum with whom he delights to be, and stands by through thick and thin in all things right and honest. We have no right, whether men or boys, to stand by anyone through thick and thin who is not in the right. Our allegiance to right is above and beyond our obligations to anything or any person on earth. Do not forget that. If the boy's chum is a thoroughly good, manly boy, the mother may feel that her boy is safe. It is not every boy, nor every farm boy, that is fit to be a farm boy's friend. There are whole classes of boys that he should avoid as friends, if he does not wish to sup sorrow sooner or later. I say "avoid as friends." I do not say avoid altogether. You are soon to go out into a world that has all sorts of people in it, from the worst to the best, and you will have to mingle with them more or less, and you should learn to touch the worse

and not be defiled. You may as well begin now and learn to be among the bad and yet not be of them.

To begin with, have the least possible to do with the boy who likes to use bad language; who loves to tell smutty stories, and who has a low opinion of women, especially of girls his own age. That is the worst sort of boy that you can have anything whatever to do with. If you like that kind of a boy, I pity you. I pity your father and your mother, and I sincerely hope you will never marry. If you choose him as your friend, you will in a few years not be fit to look a decent girl in the face. If you should afterwards repent and be converted, you will not even then be able in all your life to get rid entirely of his corrupting influence. I know men now who are trying to be Christians, and who yet, I am told, when they fall in with old chums and the like, vomit filth like a turkey buzzard. These men had filthy chums when they were boys, and they will be more or less filthy when they fall in with filthy folks as long as they live. Think what a hell it must be for a man to carry around with him filthy recollections which in his better moments he loathes and hates, and to keep on doing it until the end of his days. He had about as well be chained to a corpse. Keep your mind clean and pure and make no friendships with a filthy-minded boy.

Do not make a chum of a profane boy. He may have many good qualities, but he speaks of the God who made him in a way that even he would not allow any boy to speak of his father or mother. Either he does not believe there is a supreme being, in which case he is not a fit companion for you, or he defies Him, which is worse, or he is an ignorant fellow and uses profane language only by way of emphasis. In neither case is he fit to be your chum. You will expect to be regarded as a gentleman when you grow up, and, even if there were no sin in it, you do not want to get into the habit of using language that by common consent is never heard in the society of gentlemen and ladies.

Under no circumstances choose the bully of the neighborhood or of the school for your friend. Boys are often tempted to do so. You may admire his strength, his seeming courage, his brute force; you may think yourself safe under his protection. Do not do it. It is brain force combined with moral courage and integrity that rules the world, not self-assertion or brute force. Avoid the bully as much as possible. Do not quarrel with him; do not give an opportunity, if you can help it, for him to bully you. Keep away from him as much as possible, and when you see that your willingness to submit to his domination is regarded as an insult, get you a pair of boxing gloves and practice with your father or brothers or the hired man in the barn, and then lick the bully. Down at the bottom the bully is always a coward. I suffered much, when a boy, from this breed of cattle. My father told me that if I ever got into a fight at school, I would get a licking when I got home. I endured tortures from the bully of the school because it was known that John Wallace would not allow his boys to fight. I broke over the rule once—my father never knew it—and I had peace afterwards. Why do I insist on this point? I will tell you. The bully is a brute as a boy, a failure as a man. He develops a type of character that makes men fear him and hate him. He never has any true friends, and the man that cannot attach men to him as friends is a failure even though he be worth millions. You do not want to develop that type of character; therefore do not choose that sort of a boy as your chum.

Do not choose for your chum the boy who cannot control his temper. That sort of boy is not safe. He may have many good qualities, may mean well, but he is not safe. Solomon, the wise old fellow, saw this point long ago when he said: "Make no friendship with an angry man, and with a furious man thou shalt not go." He will make it more difficult for you to control your temper, and you never know when we will fly off and shame you. You had better be loaded down with disease or debt than with temper you cannot control.

Choose as your chum the boy that respects his father, loves his sister, fights for his little brother, adores his mother; the boy that is clean in his speech and instinctively shuns the vulgar and profane; the boy that never quarrels when it can possibly be avoided, but will not be insulted without resenting it in a manner and words, and if necessary as a last resort, by blows; the boy that is industrious, economical, and has a profound respect for things sacred. Choose as your friend the boy that has good blood in him; that comes of the best stock of people in the neighborhood. It matters little whether his father is rich or poor. Wealth should cut no figure in a boy's friendship. In point of fact, it seldom does. The only real republic that exists in this world is the republic of boyhood. This is one reason I like boys better than I do girls, and I used to like girls a good deal, and do yet. Boys do not recognize class distinctions until they become men and get spoiled. If your father is poor and you are the right sort of a boy, the best woman in the neighborhood will be glad to welcome you as her boy's chum, and if your father is rich and a wise man, he will welcome any manly boy to his home as the friend of his son. He knows the value of good blood in a boy, and by good

blood I mean that he comes from a good family whose instincts are right, who naturally like things that are honest and pure and of good report, whether they have made money or not. If you choose this sort of a boy as your chum, it matters very little whether you live in town or country. There is not much danger of your falling into bad habits. Boys of lower instincts may call you proud and stuck-up because you try to keep yourself out of the dirt. Never mind; down in their heart of hearts they respect you all the more for it.

If you want your chum to be true to you, you must be true to him. A boy that would have friends must show himself friendly, and there is a friend that sticketh closer than any brother. That friend is the right kind of a chum. That is not the way the preachers interpret this text, but it is what Solomon meant; at least so thinks Uncle Henry.

My Dear Boy:

Next to selecting a wife, the most important step you will take in the next twenty years will be the selection of the means by which you are to earn your bread and butter. It is possible this may be chosen for you. You may be the only son, or the oldest son, may be thoroughly in love with farming, and be entirely content with the lot that has been cast for you. If so, I count you happy, very happy. If you will now but read and think and keep your eyes open, your ears also, and become a thoroughly up-to-date or a little ahead-of-the-times farmer, while you may not get very rich, you will have a good chance to get as much real good out of life as any man I know.

It may be, however, that there are a good many of you, more than the farm will support; or it may be that you do not like farm-ing, or that you have the town fever. You may have neighbors and neighbors' boys who think that farmers are an oppressed people, Ishmaelites, with every man's hand against them, and you may have taken up their notions; or you may really be better fitted by nature for something else than farming. In either case, I want to have a square talk with you whether it does any good or not. To begin with, I do not think that all boys born on the farm should stay on it. There are too many of them. It will take fewer and fewer people to do the farming of the future, that is, in proportion to the population—fewer and better farmers. The towns and cities need this over-plus of the farm.

There are two kinds of boys that other professions—and when I speak of the town I mean the members of all the other professions and lines of business which for the most part live in town—can use. These kinds of classes of boys are, first, the really bright, thinking, progressive boys, strong in health, vigorous in mind, clear in thought, energetic in action, honest in purpose; and second, the young fellows who do not like the farm, who think that fortunes can be easily made in town, that town life is an easy life; who are not ambitious; who had a soft snap on their mother's breast when they came into the world, and have been looking for a soft snap ever since—born tired—possibly not their fault, who are willing to be hitched and unhitched like their father's horses. The town can use these on the streets, or in the factories and offices where the work is done by the day or hour, and but one thing is to be done, which becomes automatic after a while so that they can almost fall asleep and keep on doing it.

This last class is very apt to take the town fever. To them it seems high life, fine houses, nice lawns, lighted and paved streets, people well-dressed, working in shady offices, crowds on the streets, bands of music, pretty girls, churches, theaters, games, society, comfort. They do not know and can not be made to believe, except by experience, that every city has a White Chapel where vice reigns supreme and which no city in the world has been able to control fully, much less entirely suppress. They do not know the care-worn faces that look out of windows on the back streets, filled with failures, tailings, so to speak, that the town has hidden out of the way.

If you think, my dear boy, that town life is easier than country life, on the whole, or that it gives more average comfort, or that it has less care or requires less exertion, or that it makes better men on the average, then you are entirely mistaken. The farm boys that come to town and in ten, fifteen, or twenty years live in those fine houses and run those large establishments and shape the policies of the city and state, are of a different class of boys altogether. They are boys who learned to ride and shoot and tell the truth on the farm. The first gave them courage, the second accuracy and steadiness of purpose, and the third that integrity that lies at the basis of all success in life; in short, the qualities that make a man a success on the farm will make him a success in the city.

If you wish to choose some other profession or business, and I do not say that you should not, you should understand, first of all, that success can be won in none without indomitable energy, hard

work, and a determination to succeed that can be baffled by no difficulty, and above all things else, without integrity or character. I wish to whisper in your ear that you can acquire these things better on the farm than you can anywhere else; therefore do not be in a hurry about choosing your profession. Your constant care should be to acquire those qualities that lie at the basis of success in any business that a man ought to follow; in other words, of any honest business. When you are rooted and grounded in these, it is entirely safe to choose that business that suits your inclinations and that opens up naturally to you.

I am a great believer in providence, by which I do not mean anything supernatural or special. I believe that every boy's life is a plan of God, and that if he acquires character, integrity, or complete wholeness or soundness—that is, become what farmers call a straight up and down man—there will be an opening that will lead him into the line of business he ought to follow. I believe that if a man prepares himself by acquiring all the information possible, avoiding bad habits, bad company, and uses his time to the best advantage, matters will so shape themselves that he will find himself in the place where he of right belongs. Opportunities come right along to the man who is ready to use them. If the farm boy has acquired habits of industry, economy, truthfulness and uprightness, and his inclination leads him to be a preacher, lawyer, physician or businessman, he need not have the slightest fear of failure barring accidents and sickness, or an ill-fated marriage, if we will but take the first opening that points in the direction of his inclinations. Rest assured, however, that nothing worth having in this life ever comes without hard work, clear thinking, and right living.

There are, perhaps, some boys on the farm who imagine there is some shortcut to wealth; that dishonesty wins; that rogues prosper; and that it is little matter how you get money, or office, provided only you get it. This is about the worst mistake any boy can possibly make, and the boy who has that notion and does not get over it, can be very safely set down as a foreordained failure. There is nothing in this world that pays as large a dividend in the long run, as good old-fashioned honesty. I do not mean the corporation style of honesty. I mean old-fashioned uprightness, which is more than paying debts, and much more than telling the truth in form. It is doing the right thing at the right time, in the right way, with every man, whether friend or foe, at all times and everywhere. There are not half enough of men imbued with this kind of uprightness to meet the demands. They are wanted in every great store, factory and bank, and it is this kind of men that in the end lead in all the professions. A farm is the best place in the world to grow them; therefore, do not be in a hurry to leave the farm, and do not make a final choice of your profession or business until you are sure you are doing the right thing. If you conclude to stay on the farm and be a really up-to-date farmer, I am sure that you will get the maximum of comfort. If you choose something else and succeed, you will, in all probability, after success has been achieved, want to go back to the farm. The most of the farm boys who have the town fever and come to town and fail, would get back if they could. Mind you, I do not say that you should not leave the farm, but do not be in a hurry to make up your mind.

Uncle Henry

My Dear Boy:

I have written to you heretofore on the serious things of life, the matters that will affect directly your future usefulness, and the neglect or observance of which you will do very much to make you a failure or a success in life. I have said nothing about amusements, or, as you say, "fun," and you may, perhaps, wonder whether your father and your Uncle Henry ever had any fun when they were boys. Your father, perhaps, does not say much to you about his boyhood. He is so much concerned in looking after his farm and stock that he says little to you about the fun he had when he was a boy, thinking perhaps it beneath the dignity of a grave, middle-aged, and busy man. You sometimes wonder whether he ever had a boy's life, whether he learned to dance, to shoot or skate, or play football; or whether he went to the show or circus. You can make up your mind that he went to the show, and, after looking at the wild animals hastily, took in the circus every chance he got, and that your grandfather went with him to see that no harm happened to him, and that your father might have been seen eating gingerbread and casting sheep's eyes at one of the prettiest girls in the neighborhood, who may now perhaps be your mother. Besides, he no doubt went coon hunting in August and September, went to corn huskings, perhaps was the captain at one of these ancient contests, and was on the lookout for red ears; and if you ask him he may tell you what that means. He went to apple-butter boilings, and to wood choppings when there

was a quilting bee at the same house at the same time, and if a fiddler would even now strike up one of the old, simple melodies, I'll venture that you would notice a faraway, reminiscent look in his eyes and his feet might even keep time to the long forgotten music.

If he did not go to all these things, your Uncle Henry did, and a right good time he had, particularly when it came to getting away with the nice things with which the tables groaned in those days. There was something particularly fine about the pies and cakes, fried chicken, sweet potatoes, doughnuts, and apple dumplings. Sometimes I wonder whether women have forgotten how to make these things; and then again I wonder whether a boy's appetite does not account for the superiority of the old-time cooking. I presume the last is the correct solution. To be perfectly honest about it, I would rather go coon hunting even yet than go to baseball or football, and if I heard the well-known bark of the best coon dog in the neighborhood that showed a coon up a tree at midnight, I think I would get up at once and start after that coon a good deal more readily than I was accustomed to get up on a cold morning.

Your father, if he is the wise man I take him to be, wants you to have fun, not as the business of life, but as recreation; and your Uncle Henry regards fun, genuine, kindly fun, as essential to the boy's development as food, clothing, or education. In fact, amusement is education in the broadest, truest sense of the word; but it should be the spice of life and not the substantials; the pie and the custard after the meal, not the meal itself. There is, however, healthy and wholesome fun, and unhealthy and vicious fun. The one is life, the other is death. One develops true manhood, the other dwarfs it. The boy who learns to enjoy the right kind of fun when a boy will enjoy it all his days; and the more genuine fun he has as a boy and man as the diversion of life, the longer he is likely to live and the better his life is likely to be. I expect to have fun, or diversion, all my days, and the longer I live the better I seem to enjoy it.

Now, as to these different kinds of fun. There is no real, genuine fun in anything that is bad or vicious, nor is there any fun in anything you would be ashamed to have your mother know all about. There is no genuine fun in playing with a deck of greasy cards in the hayloft. If your mother approves of playing cards, do it in the sitting room; if she does not approve of it, do not play at all. I do not know one card from another, and I do not think I will lose anything if I never learn.

There is no genuine fun in inflicting needless pain on anything that lives. The fun that does a boy good nearly always involves

some kind of physical exercise, and with that skill of a high order. Every boy should learn to shoot, to ride, to swim, to play ball
where the games do not necessarily involve risk of life or limb, or
an undue strain on some physical organ. I do not like some features
of football; but it has, nevertheless, the essential feature of all good
outdoor games, intense energy in action. That is what we all enjoy,
whether in a horse race, dog fight, baseball or football, and the farm
boy naturally takes to amusements which require physical exercise
rather than such games as chess and billiards, which require more
delicate skill and calculation, and for the same reason that lambs
and colts and calves and even pigs play—to develop their muscles.
As we get older and the muscular system becomes fully developed,
we care less for these exciting games and take our amusement in a
quieter way.

Fun, however, is not all physical. Every farm boy should belong
to a lyceum or literary, and should cultivate by way of amusement,
not merely the intellectual side of his nature, but his sense of wit
and humor. It will be a great help to every farm boy in afterlife if
he will learn how to be a good storyteller. Storytellers, we are quite
well aware, are born and not made; so are orators and poets; but
every boy who is not totally devoid of wit and humor, can learn to
be a reasonably good storyteller if he will but study and practice.
I have always regretted that I failed to join a club while at college,
which met once a month for the sole purpose of practicing telling
first-class stories. The man who can tell a clean story that sparkles
with wit and humor is always a favorite. He is the life of every
company. It makes success as a public speaker sure to begin with,
and the ability to tell a first-class story or get off a real good joke
helps a man out of many difficulties all through his life. A farm boy
can have plenty of clean fun in learning how to tell a good story. In
fact, I know of no better way.

You will see, therefore, my dear boy, that your father and I want
you to have lots of fun. There is no reason why your life should not
have sport in it and plenty of it. You will be all the better for it as
boy and man. "All work and no play makes Jack a dull boy." Fun
is not the end of life by any means, but it makes life better worth
the living. Only let the fun be clean and wholesome, whether it be
in the line of sports which develop strength of music, steadiness of
aim, skill of hand, control of nerves, or whether it be in games that
require accurate calculation, or whether in telling stories or joke.
No right-minded boy will ever willingly listen to a vulgar, or even
half-vulgar story, no matter how funny it may be. This is absolutely

corrupting. There is no reason why a farm boy when he comes of age, should not have had twenty-one years of life as full of joy as can be experienced off the farm or on the farm in the twenty-one years that follow, provided always that he lives in a fairly good neighborhood. Even if he does not, he can get more real fun out of the colts and calves and pigs, and especially out of a first-class collie or other well-bred dog, than the town boy can get out of all the sources of amusement that are at his disposal. After all, there is no place on earth where real genuine fun can be had so cheaply and so easily as on a well-managed stock farm.

Uncle Henry

My Dear Boy:

You are, perhaps, growing restive on the farm. It has been the dream of your life to secure an education. You have envied the man who could talk well from the pulpit or platform; who could write for the newspapers; and you attributed this power to the fact that he had some time or other secured an education. You have heard it stated so often that an education is a fortune in itself, that could not be stolen or lost or burnt up, that you believe it, and think that your fortune would be made if you could secure an education. You have, perhaps, talked to your father about it and he has discouraged you. He has possibly said to you, as mine did to me over and over again, that an education would unfit you for the farm; or perhaps that he would like above all things to give you an education, but that an education is expensive, and that it is entirely beyond his power, consistent with his obligations to your brothers and sisters. You have talked with your mother about it, and she sympathizes with you, tells you she will do the best she can, perhaps cries over the fact that it is not in the power of the family to afford you this education. You have, perhaps, become discouraged over this state of affairs and concluded after all that there is nothing left for you to do but to plod along, making a living as best you can, crippled for life for want of an education which some of your chums are in a fair way to receive.

If so, you are taking a view of the subject entirely too dismal. There are three points I would like you to bear in mind: First, that there are hundreds, yes thousands, of graduates of colleges who would like to change places with you provided they had the money they have spent for their education. They would use it to make a first payment on an "eighty" and stock it and be content to be farmers all their days.

Second, that there are thousands of boys of your age that are now receiving an education, who, when they graduate, will have contracted expensive habits and will be kicked around by practical businessmen like old shoes in the street, for the simple reason that their education has not taught them to do any one thing well.

Third, that a large percent of the men who are molding and shaping the policies of the neighborhood, of the state, and of the nation, had no better opportunities for an education than lie clearly within your reach. They succeeded and you can, provided you have sufficient sand, or clear grit, to succeed.

First, I would like you to get a clear idea of what an education that is of any practical value really is. It is not something that can be poured into you as you would pour water into a bucket. A good many townspeople and some farmers talk about sending their children away to be educated, as they send the sugar box to the store to be filled. This cannot be done, no matter what time or money may be at hand. The human mind takes in knowledge as the plant takes up moisture, by free action from within, and grows, and is trained or educated by the act of appropriating knowledge. No teacher, no book, no school or college can educate you. You must educate yourself. You envy the town boy who has the opportunity of going to high school where he can learn Latin and Greek, higher mathematics, geology, botany, and all that, without either board or tuition. You say that if you had that chance you would get an education. That depends on what sort of a boy you are. The education you would get in this way, or any other, school would depend on how hungry you are for knowledge, how willing you are to apply yourself, and the natural strength of your mind. As a rule, I do not believe the town boy who graduates from the high school is any better fitted for the duties of life than the country boy who graduated from a good country school at the corner of four sections. The town boy knows more things perhaps, but the probability is that he does not know them any better and lacks the superabundant health, the keen, inquiring mind, and the practical knowledge that the farm boy must acquire on the farm if he is worth raising.

Before going any further, let me ask you if you have gotten all you can get out of the little white schoolhouse at the corner of four sections? Have you mastered the three R's—readin', 'ritin', and 'rithmetic? Are you quite sure that you are thorough master of these? Can you solve all the problems that come up on the farm? Can you measure the different fields and tell how many acres are in them as accurately as your father can, who has plowed them for the last ten or fifteen years? Can you tell how many bushels of corn there are in the different cribs? How many cubic feet of air there are in each room in the house? How many gallons the well or cistern will hold? Can you spell accurately and pronounce correctly? Can you punctuate? Can you write a legible hand, and read so as to convey to the hearer the sense of what you read? You can learn all these things at the country school, and if you can do all this, you can do more than some college graduates I know. If not, you had better take down your school books and master their contents so thoroughly that they will be like the iron in your blood.

Do not think for a moment that I undervalue an education. No one can well value it more highly than your Uncle Henry. It is the educated mind that rules the world, from the farm to the throne. I want you to have an education that will bring out the best that is in you, and I want you to get it yourself, the only way this kind of an education can ever be had; and the place to begin is with the three R's, and just where you are on the farm. If you are determined to have this sort of an education, nothing can keep you from it. "Who wills, can." Until you give up yourself you can never be beaten in anything you undertake to do that can possibly be done. If you are, I had rather not say contented, but determined to be a farmer, which I hope you are, you can be a fine, well-educated farmer without any financial aid from anybody.

Devote the next year or two to mastering thoroughly the subjects taught in your common school. Get on good terms with the teacher whether you go to school or not, and get his or her help. Put your wits to work in gathering together enough money in the next year to give you one term at the agricultural college of your state. Send for a catalogue, map out the studies you intend to pursue, and keep your mind constantly at work in that direction. If you accomplish these two things in the next year, or two years, you will have made a first-class start in the direction of getting an education, and when you go to school you will go with a first-class thirst for knowledge, with a determination to get it, and a clear idea of the value of every dollar. The boy who starts in this way will "educate" twice as fast as the boy whose father sends him to be educated with plenty of spending money.

Meanwhile, do not neglect your reading, but be careful what you read. The habit of reading worthless books is not a virtue but a vice. The habit of skimming over good books is a vice of scarcely less magnitude. The man to be prized by friend and dreaded by foe is the man who reads few books, but those of the best, and reads them so that he not merely knows all they contain, but catches their spirit.

Whether you are to be farmer or a professional man, give close attention to farm problems. One of the worst humbugs of the day is the idea that prevails among educated men that a knowledge of the dead languages is very important, if not essential, to the training, or education, of the mind. That their study gives this training is true. They act as a grindstone to sharpen the mind; but the problems on the farm, the question of the movement of the water in the soil, the structure of the plant, the methods of digestion and assimilation of food by livestock, the detection of diseases among plants and animals, and the best methods of prevention and cure, furnish the material for just as good a grindstone upon which to sharpen your mind, as the language that some people used who have been dead about two thousand years. Education, after all, is simply the fitting of the eye to see, of the hand to work, of the mind to perceive the truth, of the tongue or the pen to express it; and it is by the practice of all these that we educate ourselves and become strong, true men. You will see, therefore, that education is not a bonanza given to the rich; that it is something that can not be cornered like a grain on the market; that it is something of which the man can not be deprived who has the determination to get it, and that the share of the education which any man can receive under any circumstances is determined mainly by two things: his natural endowments and his determination to develop them. Neither money, nor schools, nor teachers, nor position, nor anything else can make a strong man out of a boy who has not the brains to begin with, or who has not the thirst for knowledge and the determination to get it. If you have these you will get the education, no matter how far off it may seem now. If you do not have them, nothing in this world can give you a real education.

It will help you a good deal if you will from time to time inquire into the history of men that are making things go about their way in state and nation. Many of these men never saw the inside of a college, of an academy, or a high school. They had no more money with which to obtain an education than you have. They do not now have what the world would call an education, but they have the real education, the development of the mind, the power that the real education gives, and that is what you are after, or should

be; and having that you have everything. What you need is to be thoroughly waked up. I have had many teachers in the course of my life, some of them drillmasters and grad-grinds who pounded a lot of knowledge into me which was of no particular use then, or even afterwards; and others who filled me with boundless enthusiasm; who set before me a high ideal intellectually and morally; and the latter are the only profitable teachers I ever had. This is precisely what I am trying to do for you. If I can thoroughly awaken you to the fact that there is but one life before you, that you must make the very best out of the talent nature has given you, must "hitch your wagon to a star" if you want to get along, I shall have done what I started out to do in writing these letters; and if you take my advice, you will thank me to your dying day.

Uncle Henry

My Dear Boy:

You have no doubt from time to time heard your father or your mother, or both, say that if they could only live to see their children well-started in life they would be entirely satisfied. For this they cheerfully toil, save and endure hardships and privations that come to all of us sooner or later. They are not so particular as to what business or profession their children may adopt. They would prefer to have them become farmers and farmer's wives, and settle somewhere near them; for to the parents the children are children long after the grandchildren come. While they would prefer, as a rule, that their children should be farmers, they will not object to one or more of them engaging in business. Many mothers would like to have one of their sons become a preacher; and every mother would like to see her daughters married to men of good character and well-to-do. This is what they mean by "a start in life."

If you keep your ears open when talking to farmers who have not succeeded well, you will frequently hear them say that the trouble with them was that they never had a "start." Other men have succeeded because they had a start. Some lay the blame for not having a start to being born poor; others to ill-health; others to a sickly wife or children; and a few are honest enough to admit that they spent their best years in sowing wild oats and are now reaping the harvest. They have evidently given up the hope of ever doing more than making some sort of a living, for the reason that they failed to get a good start at the right time. If you will get on as intimate terms as you can with men who have made success of farming, (and I advise you on general principle to do this.) some of these men may tell you how it was that they got their start while others failed. You will be surprised if you get down into the history of the lives of successful men to learn how few of them ever got a start in the way of money being given to them by relatives. You will find in almost every case that these successful men made their own start; and where they did not make their own start, they were thoroughly, and, as they thought at the time, severely trained by parents who knew from their own experience some things which I will try to tell you in this letter.

The point I wish to impress upon your mind first is that this getting a start is one of the most important things in your life. You came into this world, if I recollect right, about eighteen or twenty years ago. Time has been pushing you on right along. It has never stopped a minute for you to think what you will do. It will never stop until it pushes you through to the other world. It gives you one chance, and only one. By this I mean that you have but one life to live, and if you are to make a success in life, you must start right. If you fail in this, while you may redeem yourself to a greater or less extent, you will never be the man that your mother and father hope you will be, or that your Creator intended you to be. As time will not stop for you to think, you had better do a good deal of solid thinking while your time is going on, as to how you can make the right kind of a start in your life. You will never in all your life spend time more profitably.

In the first place, I want you to rid yourself entirely of the idea that a start in life means only the accumulation by inheritance, by gift, by trickery of one kind or another, or even by honest work, of enough money to set you up in business. More or less money is essential to a successful start in life; but after all it is not the start, even when earned by your own hands and brain. It is the evidence that the start has been made, but is not by any means the start. Most farm boys think that if they had a thousand dollars, or even five hundred, they would be well started. They say, "It takes money to make money;" and while there is some truth in this, it is not by any means the whole truth. The real start in life does not consist in what a man has, but what he is; and the value of a money start is not in the money at all, but in the qualities of the boy, mind, and heart that have been developed in making that money honestly. The boy who

is shrewd enough and dishonest enough to make a thousand dollars by the time he is twenty-five years old, by sharp practice, by overreaching, or by gambling on the Board of Trade, may think he has made a good start. Some of his neighbors may think so, too. His father and mother may be foolishly proud of him; but I want to tell you that he has made a start in the wrong way, and it were a thousand times better for him if he had saved a hundred dollars in that time by honest work and careful economy. In the last forty years your Uncle Henry has seen a good many young men make that kind of a start, and today he can not think of one of them who is either scratching a poor man's head or has failed to retain the confidence of those who knew him.

If you are to succeed you must get a start of the right kind, and you cannot get that without a good deal of hard work, close economy, and more or less of self-sacrifice. You will have to work, and you will have to think, and you will have to do without a good many things which at first blush you would like to have. I will first tell you about some ways in which you will not get a start. You will not get the right kind of a start by going in debt for a courting buggy, to spend your evenings in going to dances, circuses, etc, with some good-looking girl, who, if she would speak out, does not value you above one of her hairpins, who eats your caramels and ice cream, thinking, if she thinks about you at all, that you are a silly goose for wasting you substance in that kind of entertainment. She more than half suspects that the buggy is not paid for, she knows you are wearing more stylish clothes than you can afford, and she secretly makes up her mind that while she will have all the fun she can with you, she will say "yes" to an entirely different sort of fellow.

You will not get a start in life by forming the bad habit of smoking or chewing, or drinking beer and an occasional glass of whisky, nor by having "a high old time" when you go to Chicago with a carload of your father's cattle.

You will not get a very good start in life by imagining that, being raised on the farm, you therefore know all about farming, concluding that books and papers that discuss farm problems are not worth your notice.

No matter what business you may choose, there are three or four things you must have if you are to start right in life. You must have a capacity for steady, endless, hard work. There is no honest business or profession in life in which this is not a prime requisite and an absolute condition of success.

You must think as well as work. It takes more than hard work to win. It is hard, intelligent work, where the thinking brain guides the hand, working according to a well-defined purpose. My father used to say to me: "Henry, if you don't think, it makes very little difference whether you work or not." That was sound advice fifty years ago; it is sounder advice today than it was then.

Getting a start in life means being absolutely honest. I do not mean by honesty merely the willingness to pay debts. That is a part of honesty, but a small part. I mean uprightness, integrity, reliability, truthfulness. I mean that quality embraced in all these words that will lead your neighbors and all who have any business with you to rely absolutely on you, with the utmost confidence that you will do what you say you will do, and that you can be depended upon under any circumstances; or, in the expressive language of the farm, "will do to tie to."

Now, the value of the first thousand dollars you may earn, is not in the money, but in the training that making it in an honest way will give you along these lines. Therefore, start out to make this thousand dollars, commencing with five, ten, fifty, one hundred, or five hundred dollars earned by yourself, by your own unaided efforts, digging it out as the miner digs the gold out of the Klondike. There will not be so much trouble in the making of it as in the saving, in avoiding the spending of it for useless things, and in putting it at interest as fast as made; or better still, investing it in young stock to which you will give your personal care, and thus learn to feed and breed, to buy and sell. It is astonishing how fast a young man on the farm who has a kind and wise father can accumulate money in this way, and with it the qualities of mind and heart and hand that make the real start in life, no matter what the future profession may be. When these qualities have once been acquired, the amount of money that has been accumulated in acquiring them is really a secondary matter. Any farm boy who has been wellborn— by wellborn I mean has come of good, honest, respectable parents, whether they have much of this world's good or little—who has fairly good health, and such education as a common school can give, can acquire them if he will; and if he does not, he has no one to blame but himself. He must, however, bend every energy to its requisition. He should not think too much about the girls until he has made a start. The good ones of them will keep. He must avoid acquiring expensive habits, and diligently school himself to hard work, clear thinking and honest living.

In every department of life, whether manufacturing, merchandising, or railroading, the patrons of every profession are looking for boys who have that kind of a start. If you go into any city, large or small, in this broad land, you will find that men who are running things had that kind of start, and got it themselves. Boys who have that kind of a start do not need to do much advertising. True manhood has a ring to it which all worthy men recognize, and that ring can not be counterfeited successfully. There is no place where a start of that kind can be obtained as well as on the farm. Having secured the money part of the start, the farm boy can spend it in obtaining an education in college, or in the particular branch of business, or the particular profession which he chooses to follow; and barring sickness, accident, or an unwise marriage, nothing can prevent him from making a success in life. He may not become a millionaire. A few, but only a few, really honest men do. He may not rise to high political position, and yet he may; for after all, honesty is the best politics, although few professional politicians seem to think so; but he will secure the confidence of all who have to do with him, all the world's goods that he really needs, and more too; and when time pushes him out of the world, as it does all of us, he will have left the world a good deal better than he found it, which at last is the highest measure of success.

This is the kind of man I wish you to become. It is in your power, when you start out resolutely, to make this start, or live to regret in after years that you did not take the advice of your Uncle Henry.

My Dear Boy:

When you and I were quite small we had a great difficulty learning to walk. We first crept, then with great effort we learned to stand alone by a chair, then to take a single step, then two or three in succession. Our mothers encouraged us to make longer ventures, and by and by we learned to walk across the floor, falling down, perhaps, two or three times; and when we succeeded we felt that it was the proudest day of our lives. Every step at the beginning required a distinct effort of the will, of which we were then conscious, but it was not long until we walked without conscious thought or movement. In other words, it walked itself. We had acquired the habit of walking.

It took us a long time and much conscious effort to learn to
talk. First, we managed the easy words of one or two syllables, then
of three, and by and by, with much toil and pains, we managed the
big words, and the w's, v's, and the h's. We were quite proud of
ourselves, and our parents were prouder still of us, when we learned
to talk, or rather, when it learned to talk itself; and the only trouble
they had for some years afterwards was that we talked entirely too
much. We have formed the habit of talking.

It took you and me a long time to learn to read. We had first to
learn the name of one letter, and then of another, with their appro-
priate sounds, and then combine the letters and sounds, so that we
did well when we could make out one word at a time. We forged the
habit of doing this, and now we can read faster than we can make
the sounds. Some people have learned, not merely to take in a word
at a time, but a sentence; and can skim over the pages of a book,
and get the sense of it, in a way that those who have not learned to
do so can scarcely understand. At least, I can not. They have formed
the habit.

When you were in my office last, you noticed how very rapidly
the stenographer handled the keys of the machine. It would be slow
work for you and me, but if we had formed the habit it would do
itself. It was slow work for the stenographer to learn to take down
talk in shorthand, but it became so easy for a man I once employed
that it required no thought at all; and on a hot day he would go
to sleep taking it down, and I used to have to waken him. He had
formed the habit.

For several months of my life I had to taken down speeches and
lectures in longhand, and I got in to the habit of leaving out nearly
all the vowels in writing, and part of the consonants. Since then
I have written a hand that few can read. I get to thinking, and the
pen wiggles—that is all; and often I cannot read it myself, unless I
know what I am writing about—a very bad habit.

I have given you these illustrations of the power and force of
habit for a distinct purpose. You and I are simply bundles of habits.
Every time we do anything it becomes easier to do it in the future,
until, by and by, the doing of it becomes unconscious, automatic—it
does itself. It is, therefore, of the utmost importance that we correct
habits in doing, in thinking, in living. If we learn to do a thing badly
and form the habit, we will in all probability do it all badly all our
days. If we form the habit of doing things wrong, and if we form the
habit of doing bad things, we will in time become bad men; for bad-
ness and goodness are, to a certain extent, at least, matters of habit.

When your Uncle Henry was a boy, he was very anxious to get over a great deal of work. For instance, he was anxious to be the fastest corn husker and the fastest grain-binder in the neighborhood. Unfortunately, he formed the habit of binding sheaves loosely and failed to acquire the habit of getting all the silk and husks off the corn. The mice had a picnic in the corn that he husked. A loose sheaf when hauled in, or out at threshing time, was instantly recognized as one of "Henry's sheaves." I tried hard to correct this habit in after years, but never succeeded. I could bind tightly enough as long as I kept thinking about it; but the moment I began thinking about something else, and that was about all the time, the sheaf bound itself loose.

You will avoid a great deal of trouble in life after if you will acquire the habit of doing whatever you do well. It takes no longer to acquire the habit of doing it right than wrong, and when a habit is once formed, it stays formed; and the longer you practice it, the more firmly the habit becomes fixed. It is just as easy to curry the horse well, when you get in the habit of it, as it is to give him a "lick and a promise." It is just as easy to milk the cow clean, and with neatness and dispatch, as it is to milk her otherwise; and the habit once formed of doing things right will stay with you as a perpetual heritage and blessing. The habit of doing things right can be formed by the conscious doing of them in the first place, and every subsequent repetition of the act fixes and confirms the habit until it becomes the permanent habit of life. The man who learns to do the work on the farm right will be very likely to do all his work right, for the reason that it becomes second nature.

I need to say to you that you are very foolish if you acquire what we ordinary called "bad habits," which are usually regarded as habits of doing wrong or useless things, and not the habit of doing right and useful things in the wrong way. A boy is foolish to acquire habits which involve expense or injury to his health or waste his time and money. There are weights enough to be borne in life without taking on extra loads and binding them to our backs by the silken, yet unbreakable chords of habit.

Mental and moral habits are even more important than physical. You will be greatly helped in forming right moral habits, and continuing therein, if your father and mother have in your earliest childhood thoroughly instructed you in the first principles of right and wrong; have taught you to do right because it is eternally right, and taught you to avoid doing wrong because it is everlastingly wrong; and that while there may be palliation of the guilt of wrongdoing, there can never, under any circumstances, be a good excuse for it. If you are fortunate enough to have this kind of early teaching, and take to it, it will be comparatively easy to form right mental habits.

You will be much more likely to adopt these elementary principles of righteousness if you have the righteous blood behind you, and unlikely if you have bad blood coursing in your veins. For, though some affect not to believe it, it is a truth as old as Moses, and in fact, very much older, that the iniquities of the fathers (and mothers, too) are visited upon the children to the third and fourth generation of them that hate the Lord, and mercy shown to a thousand generations of them that love Him and keep His commandments.

Some good people do not like this text because they do not understand it. It is to me of great interest to know that in three or four generations inherited evil may be overcome, and that when overcome, a righteous character may be perpetuated under the right kind of training for an indefinite period.

You will readily understand, therefore, that the formation of mental and moral habits is about the most important thing in your life. For example, you may form the habit of seeing things clearly and distinctly and stating them truthfully; or you may form the habit of half seeing, and stating them loosely. Do you know that as a matter of fact there are comparatively few persons that can tell the truth, that is, state things precisely as they are? They are not conscious liars, not liars at all in the obnoxious sense, but nevertheless we cannot depend upon what they say, because we know they are not in the habit of seeing things as they are, or of stating in exact language what they do see. There is no habit of more value to a young man, whether on the farm or off it, than to be able to discern truth, fact, reality, and to state it as he sees it without exaggeration or being influenced by his hopes, his fears, his dislikes, or his prejudices. The farm boy has a better chance to practice in this line than any other class of boys. He should train himself to know by observation the weight of the steer or calf, the size of the field, the distance from one point to another, the yield per acre of the crops grown, and the color and form of every particular animal on the farm. The shepherd, by his close powers of observation, can tell each particular sheep if there are five hundred in the flock and detect it at a glance over the flock which one is missing. The farm boy can do the same if he but will.

Next to the habit of seeing things as they are, is stating them, or telling the truth, a habit that can be acquired only by careful and long-continued practice, and which, when once acquired, will do

more to win the confidence of men than almost any other one trait of character.

Still more important, if anything, is the cultivation of the habit of right-doing. It is almost as easy to do right as wrong, if one but acquire the habit of it. Habits of right feeling precede habits of right-doing. The thoroughly good man does right without thinking about it, or talking about it, or taking any credit whatever for it to himself, because he has formed the habit of it. It becomes part of his nature. It would hurt him to do anything else, because it is the breaking up of the habits of his life, a sort of rupture of the fibers of his being.

If you will think a moment, you will see that it would be possible for this world to run smoothly in no other way than by making men bundles of habits, thus giving permanency to character. You know that your father will be about the same sort of a man tomorrow, next week, or next year, that he was yesterday, last week, or last year; that there will be little or no change in his walk, in his talking, in his modes of thinking, and manner of meeting whatever problems come to the farm. He will be out of humor with the same sort of things, and be pleased with the same sort. He will like the same sort of people he has liked for years, and will dislike the same sort. It will be the same with your mother, and with all your neighbors. Businessmen, politicians, preachers, teachers, all who have to do with men, count on this general permanency of character, the results of fixed habits based on fixed principles. If it were otherwise, it would be impossible to do business, or to get along at all comfortably with each other; and society, politics, church affairs, and everything else would be in total confusion.

Getting acquainted with men is simply taking stock of their habits, and we are greatly surprised when some friend develops a trait of character which we never saw before, because we had never become acquainted with that particular habit in the bundle. It is, therefore, of the utmost importance that you form the right sort of habits, and do it when you can. The older you become, the harder it will be to form good habits and to break up bad habits.

You are yet young. You can form the habit of doing that which your own conscience tell you is the right thing to do; that for which there is no necessity of making any excuse; that of which your mother and father approve, and of which your own sense of right approves, and of which the rule of this world approves. It is not easy to do this as to do the other thing, for there is more or less weakness and inherent wickedness in the best of men; but the constant doing of it will so fix the habit that when you go out to take your place in the world, you will never seriously think of doing anything else, nor will the world expect anything else of you. It is in this way that men become strong and form characters on which the weak lean for guidance and direction. If you are in doubt about the propriety of doing anything, do not do it. "He that doubteth is damned"; that is, condemned, or reproved, by his own sense of right or conscience.

You will see, my dear boy, that I have not given you a lecture, nor preached a sermon, but simply pointed out certain facts that you should know. It is not a question as to whether you will form habits or not. Form habits you will; you can not help that. It is simply a question whether you will form right habits or wrong ones; which means, whether you will be a man that "will do to tie to," or not; whether, in short, this life, the only one you have to live in this world, is to be a success or a failure.

Uncle Henry

My Dear Boy:

I do not wonder that you sometimes become restless and want to get away from the farm for a day or two. I, too, felt that way. From the beginning of spring wheat sowing to the end of corn husking, is a long time for a boy to work hard, day and night, with no vacations except the Fourth of July.

Time flies after a man has passed fifty, but it limps along very slowly with a boy under twenty. Work on the farm, although much easier than it was when I was a boy, is not after all the very easiest kind of work, and if continued right along without interruption, becomes very monotonous. The hours of work are long in the summer, the night shorts. We get tired looking day after day, and month after month, at the same horizon, which seems to close down on us and shut us in all around, knowing all the while that there is a great world beyond, throbbing and palpitating with human hopes and ambitions. We long to see something of it and share in its abundant life. At least I did. I know you do.

Valuable as is the drill of farm work in forming habits of steady, persistent industry, the boy needs once in a while to get away from home, to see something of what the papers tell him about, and to measure himself with other boys, and be measured by them. If a bright boy, he is apt, if kept always at home, to become a conceited fellow with vast conceptions of his abilities in one direction or another, and, if he is to be of any account in the world, needs to have this conceit completely knocked out of him. The farm boy who is first in the common school is very likely to get what, in common parlance, we expressively term "the big head," and should have a chance to meet someone who has forgotten more than he ever knew. The neighborhood bully should be encouraged to meet someone who will take the swagger and insolence out of him with one swift blow coming like a clap of thunder out of a clear sky. The boy who has fallen into careless habits of speech or behavior should have a chance to see how well-bred boys conduct themselves, while the modest and diffident boy should have a chance to learn that an honest heart and a clear head with country manners are like gold— current the world over at full weight, with or without polish.

The behavior of the farm boy away from home furnishes an excellent means of judging what sort of a boy he is, what sort of folks his parents are, and in what kind of neighborhood he was brought up, or at least, the kind of associates he has. When I see a number of farm boys going home from the state fair or any other public gathering—noisy, profane, and evidently aiming to attract public attention—I am not surprised if I notice a bottle of liquor circulating among them, and I infer that they have seen but little of the world, and that little not by any means the best part of it. I expect, of course, that when a farm boy goes away from home he will be somewhat like a colt that has been kept in a stable and needs exercise. I expect him to have a good time and to enjoy it; but I also know that, now that he is off his guard, I can form a good deal better judgment of what he is, and what he is likely to be, than if I met him on the farm, and under his parents' eye. I like the boy that likes fun. I like it myself, and better in my old days when I was young; but there is no real fun in any behavior that is loud; that has neither wit nor humor in it, but more or less obscenity and profanity. I like to see farm boys, when away from home, take an interest in baseball and football, and take part in these games, if they are stout enough to do it without danger. Such fun is natural, and as healthy as for colts to run, or for lambs to play. A boy that is fair in games will likely be fair in business, and conversely. I have always suspected the genuineness of the Christianity of a certain preacher who once tried to cheat me in playing croquet. The boy, who, when away from home, wants to see the seamy side of the city, or to "paint the town red," or to have what he calls a "high old time," serves notice on all men that he has poor stuff in him, and is likely to make a poor use of it.

Other farm boys when away from home reveal the fact that they are insufferably vain and conceited, and need to be taken down severely a peg or two. These are not bad boys. They simply overestimate their good looks, or their smartness, or, perhaps, their father's wealth or social position. Living in the narrow circle of the neighborhood, they get an enormously exaggerated idea of their own importance and make themselves the laughing stock of sensible people. If they have sense enough to see this and get down from their pedestal, they will come out all right; and it is often an excellent thing for a boy to get away from home and be laughed at and ridiculed and made to feel cheap and mean. The medicine is hard to take, but it is good for you. You will not get it unless you need it. The next time you get away you will not dress nor act so as to attract attention. You will slip along quietly like the rest of us common folks, and will, as a result, get the goodwill of the plain, common sense sort of people who are really the sort that can be of any help to you. Let me give you a hint: Whether away from home or at home, dress, and act so that you will attract as little notice as possible. Leave off that glaring necktie and the hat that is either too broad or too narrow in the rim. Do not push yourself into public notice, and do not hide away. Face the world boldly and modestly, do not force your own opinions upon people, and do not hesitate to express them modestly, but firmly, when called upon, and the outside world will henceforth consider you a boy of good sense, and the making of a strong man.

A good many farm boys are entirely too modest and diffident when away from home, and particularly so if they are thrown among noted men, or men and women who have seen much of the world. They then become painfully self-conscious. Their dress does not seem to fit as they thought it did when they left home. They lose their natural manner and become stiff and awkward, especially when in the society of refined and cultivated ladies. They are at a loss as to what to do with their hands and feet. This is a very painful experience. Do not fret because you step high while other men seem to glide along. You are accustomed to walking over rough surfaces; they over carpets. Do not feel badly because you speak loudly. You have to speak loudly out of doors on the farm; they speak to people in the house. The sensible man understands all this and thinks none the less of the boy for acting naturally and farm-like.

Do you know that businessmen of all sorts are constantly looking out for just this sort of boys? They always suspect the farm boy, who, when away from home, tries to ape the manners of the town boy, or who shows traces of foppery. In the eyes of sensible men in the city, country manners are always at a premium. You may not know, but I do, that successful men everywhere like well-bred, modest boys, and will always encourage and push them to the front as far and fast as it is safe. Most of them were farm boys themselves, remember their own early trials, and take genuine pleasure in giving a helping hand to these young fellows who come to the city to push their fortunes, or who push them on the farm.

I can point out middle-aged bankers who will loan money to a boy, when, if the father came, they would have no money to lend. They see clear through country manners to the real stuff underneath, and take pleasure in helping the farm boy with a clean life, resolute will, and unstained honor. There is one man now gone over to the other world, whose memory I revered. He invited me with other boys to take tea and spend evenings. His wife was city bred, stylish, and vain. I was fresh from the farm, awkward, and very plainly dressed. She lectured me on my lack of taste in dress and refinement in manner. He overheard it and said: "Henry, you will find it much easier to put her advice in practice if you stick closely to your studies, get the foundation first, and be thorough in all your work. Your dress and manners then will come all right. Make yourself worth polishing, and the polish will come as you rub against men."

While self-conceit and self-assertion should be repressed in the farm boy, he should at the same time know the full value of his powers and learn to rely on them. Getting away from home and mingling with the very best sort of people will teach you how to take your measure. Low-bred fellows, physical and intellectual bullies, and small souls who are constantly in fear lest some one surpasses them, will try to intimidate you by ridiculing, by browbeating and bulldozing you; but when you strike a true man, he will be your friend. It will do you no good to associate with snobs and upstarts. If you get a good, honest, manly, and intelligent face on you, which you can get only by being an honest, manly, and thoughtful farm boy, you do not need any certificate of character or letter of recommendation from anybody. You will find that the very highest people in the whole land are the most easily approached, and the most ready and willing to help a modest, thoroughly upright and self-reliant farm boy. I have the honor to be personally acquainted with many of the great men of the nation, and I find that the greater the man, the easier it is to approach him. It was much easier to reach General Grant during the war than the petty officers who waited on him. President McKinley is a much more approachable man than many of the little popinjays who want to be county officers, or to be elected to the legislature. You would feel much easier in talking with the members of his cabinet than with many of the little snipes who hold clerkships in the various departments.

Get away from home if you can, and when away mix with the very best people within your reach. Keep out of the noisy, boisterous crowd. Let the prigs admire their own excellencies. Do not hesitate to mix with the best men much older than yourself. You need never be afraid or ill at ease with a really great man. If you are of the right sort to begin with, they will be glad to talk with you. If the best men give you the cold shoulder, there must be something wrong with you. What is it?

Uncle Henry

My Dear Hobbledehoy:

You don't know me, but I know you; for I was a hobbledehoy myself, and have lived through it; but the memory of the joys and sorrows of that time has survived the lapse of more than sixty years.

You have great joy of a good appetite and good digestion. Oatmeal, ham and eggs, followed by buckwheat cakes with real maple syrup or some right good sorghum molasses, bread and butter, real coffee with genuine cream, all disappear from the region of your plate like snowflakes on a warm spring morning. Or, if it should not happen to be the right morning for a meal like that, then flannel cakes, fried potatoes, side-meat, bread and butter, and milk will disappear in a like manner. Something within you makes some marvelous disposition of it before noon; and you are ready for any square meal that the season may afford. By suppertime you are hungry as ever; and your sister, when in a bad humor (if ever she is), may make some remarks about a glutton, which you are not. You are merely a hobbledehoy, who is growing, and growing fast, and needs good solid food, and plenty of it. You are keeping the active machinery going, and adding to it every day; and hence you need both the "food of support" and the "food of increase," as teachers of

feeding say, or, as the farmers say, the "food for maintenance" and the "food for growth." My how you are growing; and if you don't eat twice as much as your mother, and fully half as much as your father, you are either not living up to your opportunities and needs as a hobbledehoy, or they are eating too much for their own good.

It is great, as I remember it, to have a vigorous body, with perfect digestion and assimilation; to be an incarnate appetite, and to have no inward pains and no discomfort unless you gorge on green apples. Then you only get what is coming to you, no more. Take a little advice from an old-time hobbledehoy who has long been gray-headed: If there is anything you particularly like, don't eat too much of it at any one time. When a hobbledehoy, I was very fond of ripe red plums of the European sorts. I transgressed. My "Little Mary," (a pet name I have for my stomach) rebelled, and to this day protests at the sight of plums. I have lost much satisfaction in life because of that one transgression. Another day, when very hungry, I ate too many cookies with caraway seed in them; and to this day something within me rises against the taste of caraway, no matter what it may flavor. Therefore, be moderate in eating things you especially like. Wise old Solomon once said: "Put a knife to thy throat, if thou be a man given to appetite."

You have also, in a way you will never have it in later years, the full joy of living, just merely living—to fill your lungs with fresh air, to stretch your muscles, to run and jump and play as do the lambs and colts and calves, to feel the thrill of the first rays of the sun shining through the haze at dawn, that peculiar atmospheric wine, which the old prophet felt to be "the hiding of the Divine power." It is great to imbibe this ozone at any time of life, but never so great in later years as when you are a hobbledehoy or the sister of one.

As a hobbledehoy you really joy in doing things worthwhile. Your sister, and possibly your father and mother, sometimes speak of you as lazy, especially when it comes to chopping wood for the cookstove, or carrying water during the noon hour, when the hired man is taking a nap "under the shade of the old apple tree" on a hot day. They are pretty sure to think you are lazy, if they have asked you to work in the garden, with more weeds and harder clods than you ever saw in the cornfield. I know my parents had just such an opinion of me. How I did hate to work in the garden and kill weeds with the hoe instead of killing weeds in the field wholesale with the harrow, and mashing clods with the big roller.

If your parents explain to you why this work needs to be done, and show you that you found it really worthwhile, and you are allowed a little discretion as to the way of doing it, they will not find you lazy. If your father will give you a colt or a calf, even a runt pig, on condition that you will raise it and care for it, they may be sure that they will not find you lazy in looking after it.

If they will talk about "our farm," "our" stock, and "our affairs generally"; if they will develop in you a sense of partnership and pride in achievement; if they will show you that your part of the work is a part worthwhile, and that you are in every deed and truth a trusted member of the family, they will find that you have a great joy in achievement. If not, there is something wrong with you.

You have your sorrows as well as your joys, my dear hobbledehoy. At least, I had, and I have not forgotten them. It is not a great sorrow, but it is an unpleasant thing, after you have carried in coal and wood and water, to have your sister growl at you because your boots were muddy and tracked up her clean floor. You would not have minded it so much if she had gone at it in a little different way. In fact, you would then have felt a bit sorry about it, and would try to do better; but you don't like to be growled at or scolded. You may feel hurt if your sister has the best room in the house, and always in apple-pie order, while any old room, kept in any old way, is thought good enough for you. You don't like it when your brothers and sisters make fun of your awkwardness—for you are awkward and can't help it—or make sport of your budding mustache, and tell you that your whisker-seed was not the right sort or color, and suggest that you put cream on your face and have the cat lick it off. You don't like to have these youngsters make fun of your necktie, or your stand-up collar, or the shine on your shoes. You think it would not be dignified to scold them, and you can't very well spank them.

They will have their fun at your expense, and all you can do is to grin and bear it, for you are only a hobbledehoy—neither man nor boy.

You can bear all this, but you can't bear to be scolded when you don't deserve it, scolded or blamed for something you did not do, or for some accident which was not your fault; that hurts. You would not admit it so much if you were really a fault. You can take that as something coming to you, as you did a deserved whipping when you were a kid. You really felt better after it was over with; you felt that justice had been satisfied, and you had thus made atonement.

But to be treated unfairly by your parents, or to be suspected when you had done nothing to deserve it, that is a real sorrow, and you don't get over it very soon.

To have done your very best at some new task, something you never did before, and then to be scolded for not doing it better; or to have your best efforts ridiculed and laughed at, as sometimes happens even in good families; that hurts. You know how it does; so do your parents, who are afterwards sorry, though they do not always have the courage to say so, and thus salve the wound inflicted thoughtlessly.

You are a growing boy, but sometimes you have a man's sense of justice; and nothing so rankles in your mind as a sense of being unjustly treated. For instance, if your father gave you a calf, which you have cared for faithfully, and at last it proved to be father's steer, and your work and ownership were forgotten; that hurts you most, in that you can not think quite as much of your father as you did before; can not, if it were to save your life. If money was badly needed at the time, and your father had told you about it, and had asked your consent (which, if you are as much of a man as I think you are, would have been given freely), you would not have minded it very much. If he had given you his note for the price, with interest, you would have been supremely happy. This he should have done.

But, my dear hobbledehoy, these are all trifles, even if they seem to be and are real sorrows now. These, with the pangs of puppy love which you are likely to suffer about this time, and jealousy of the fellow who cut you out, who seems to be in high favor with her just now, and whose head you would like to punch, will all pass away with the coming years. If you have met these sorrows bravely, you will be all the better man because of them. After a while, you will laugh at what seem to you now real sorrows. If, however, you get sour and nurse your grievances to keep them warm, and resolve you will never forgive, but you will get even in some way, then let me tell you right now that you will probably never be much of a man.

Uncle Henry

Hardups, Hardmans, and Richmans: Cautionary Tales for the Farm Hobbledehoy

My Dear Boy:

I am quite sure that somewhere about your father's farm there is a scrap heap on which is thrown from year to year a miscellaneous assortment of stuff which is thought to be of no further use. [138] It may be an old mower or hay rake, broken plowshares or moldboards, an old shovel or spade, pieces of broken boards or of rotten posts—any old thing that your father wants to get out of the way. He may be one of these wise old farmers who manages to get rid of some things before they are quite ready for the scrap heap. He has a plow that will turn a furrow, but takes too much horsepower and does not do a good job; or a mower that seems ready to break down on the slightest excuse, or a wagon that has seen its best days. He gathers up this stuff and takes it to the nearest farm sale. Here he gets rid of it to some man who likes to buy such stuff, provided he gets it cheap, or thinks he does. There are such men in almost every community. In this, your father is a wise man, a wiser man than the one who buys this stuff; for he wants work out of his tools, and will have only tools that can do the work.

There is a scrap heap somewhere around every factory. The wise manufacturer wants service and efficiency in his machines. He must have it, or he can't compete with men who do have this efficiency. Therefore, when he finds a machine that will do the work as well with greater speed at less expense, he scraps the old machine and buys the new. Generally, like your father, he sells the old one to some man who does not know his business so well, or perhaps is not able to buy a better one; or perhaps he trades it in on the new one of improved type and gets what he can for it. For example, *Wallaces' Farmer* has scrapped two or three printing presses in the last nineteen years, and will soon have to scrap one which cost us twenty thousand dollars, because our subscription list is getting too big for it to handle.

Society has its scrap heap, not dead ones such as we find on the farm and in the factory, but living scrap heaps, into which it remorselessly drops the inefficient and the unfit. You will find such scrap heaps in the slums of the city or town, away back somewhere

where rents are cheap, and houses poor and badly lighted and ventilated, and where vice reigns and children go barefooted in the cold.

Why do people live there? Some because they can not afford to live anywhere else. They have failed. Others because they prefer to live among people of low tastes and loose morals. Why? Because "birds of a feather flock together." Therefore they have been mercilessly scrapped by the business and the social world.

My eyes are not as good as they once were, and after I am tired reading by artificial light, I got to bed and lie there and think for an hour or two. My thoughts often turn to the human scrap heap, and I wonder if many of the boys and girls who read *Wallaces' Farmer* will find their way there. I wonder what I can say to them that will help them keep out of it and enable them to stand the tremendous wear and tear of modern life, whether they stay on the farm or seek their fortune out in the wide world.

Some of you boys, I fear, will be scrapped on account of lack of good health. I pity you, especially if it is not your own fault. Some of you may inherit a weak constitution; some, but not many. In case you do, if you can overcome that anywhere, it is in the country, with its pure air, its clean living, its abundant sunshine, its pure food, and healthful exercise connected with the work of the farm.

Some of you, I fear will be scrapped for lack of ginger, the get-up-and-go that enables even a small delicate boy to do the things which many of the strong have done. Grow ginger. It's one of the best crops on the farm. It is not absolutely true that "who wills, can"; for accident, unavoidable sickness, or the burden of sick folks often prevent it; but it is absolutely true that he who does not will or determine to do, can not. Therefore, if you don't want to go to the scrap heap in the country, or drop out of sight in the city slums, you must aim at doing something worthwhile. If you won't, you can't. There are no two ways about that.

Many farm boys will be thrown into the scrap heap because they don't use their brains on the farm. They get the idea that the boy has no chance on the farm. They think that farm work is drudgery, and go at it in the spirit of the drudge, like a slave to his toil. They imagine that if they were only in the city, they would do big things. Now, boys, get that kind of nonsense out of your heads. The boy that does things on the farm, that puts brains into his work there, does things when he goes into town. The boy that learns to be a drudge on the farm probably will be a drudge in town, and is in danger of going into the scrap heap.

Many farm boys will go into the scrap heap because they form bad habits as soon as they come to town. They think it manly to drink or swear or use vulgar language, because they hear some town men do so. They think they must "sow their wild oats" in order to be manly. Boys, don't be fools. It is not manly to do any of these things. It is utter folly. The men who do the big things in town are most of them clean, strong, men (and the few rough men who do succeed for a time, do it not only on account of their roughness, but in spite of it); and they are looking for clean, strong, country-bred boys to help them. They will not trust a young man who is known to drink. They are suspicious of him if they know that he drinks even occasionally. They distrust a man, young or old, who swears. They despise a man who is foul-mouthed and relishes a dirty story.

If you boys, when you go to town, if you do go, are clean in person and in speech, if you have control of yourselves, and have the country virtues, there is not much danger of your ever going to the scrap heap. If you don't have these country virtues, you can't stand the strain of city life, and down you go. Once started on the downward path, you are likely to go on till you reach the social scrap heap.

Above all things, keep yourself clean morally. Think of your mother, your sister, the little girl you thought so much of in school, and live worthy of them. If you don't, you are in great danger of going into the scrap heap. There are plenty of good, clean men in the city, men who do things, and who are looking for good, clean, strong, bright country boys—who have not been enervated by luxury, who are energetic, industrious and trustworthy—to help them in their business, to try them out; and sometimes to try them pretty severely, for they want to find out just what is in them. If these boys prove worthy, they will advance them; but if unworthy, they naturally and wisely reject them. Once rejected, these boys are in danger of going to the scrap heap.

The way to win in the city is first to prove yourself fit in the country, in the home, and in the neighborhood. When you go to town, go to your father's church, or, at any rate, go to some church, and go to some Sabbath school. Keep out of the saloons and all sorts of bad company, if you wish to retain the confidence of men worthwhile, if for no other reason, because big men are looking for bright boys, look for them in good company and in some church or Sabbath school, whether they themselves go to church or not.

In reading over what I have written, it occurs to me that you might think that the great object of the farm boy is to fit himself for town life, and to keep out of the scrap heap after he gets to the city. On the other hand, I think that country life is the ideal life for which

country-born men who have had experiences in the city ardently long.

What you need is to fit yourself for life in the country; and the best equipment for those who find themselves better fitted for city life is the training which the country life gives. All the boys can not stay in the country. Some from necessity, some from fitness, some from inclination, will gravitate to the city. The boy who does not develop a clean life, a vigorous health, and in general, the country virtues, will not succeed in the city if he goes there, and will be in great danger of landing in the scrap heap. The primary conditions that make for success in life are very much the same the world over, and fortunate is the boy who is born on the farm, of good parents.

My Dear Boy:

I fear I have given you more good advice in previous letters than you will take. I suspect you have not read all of them very carefully. You may have the idea that it is natural and right for a boy to have his fun, and perhaps sow a little wild oats, and have a good time generally until he is married, and then will be the time to settle down. Your father and mother have read these letters, and perhaps urged you to read them carefully. They may, indeed, have urged you a little more than was prudent. I find nothing does a boy good unless he relished it; and unless you have a taste for good advice, and take to it naturally, it will, very likely, be wasted. For this reason I propose hereafter to give you illustrations rather than advice, and will tell you something about farm boys who have failed, and failed for the reason that they have not taken advice similar to that which I have given you.

When a man has lived over sixty years, and been a close observer of farm boys, mingling with them for nearly fifty of these years, he acquires a very large acquaintance, and can group his acquaintances into a great many large classes. I propose to tell you this time about a very large and respectable class of farm boys under the general name Hardups.

The Hardups are a very old family, their pedigree tracking back through the revolutionary period, and quite a number of them came over in the Mayflower. In my trips abroad I find that they are a large family on the other side, and in looking up their pedigree, I have found that they antedate the oldest names in English peerage, and in fact trace to that period "of which the memory of man runneth not back to the contrary."

A large number of them came West before and after I did. I have kept my eye on a few, and have had many pleasant and profitable talks with them. Many of them are eminently respectable people, and some of them are among my warmest personal friends, of whom I may speak freely, provided I don't tell where they live.

For example, there is my friend Ben Hardup, as good a fellow as ever lived, true to his friends, openhearted, generous, loyal to his party, devoted to his church, and true to his name, Hardup. I knew his father before him, his brothers and sisters, his cousins and his aunts, and they were all of a piece—"the easygoing Hardups," we used to call them. They were good livers. I shall never forget a remark that Ben's father made to me one time at supper when I asked him how it was that he was able to live so much better than many of his neighbors. He was carving a fat turkey at the time, and he stopped, looking at me with mock severity, and said, "Henry, I want you to understand that neither I nor any of my boys will ever die in debt to his stomach." That particular branch of the family, at least, never did, whenever I have known them.

I warned Ben when he came West not to settle in the timber, as I have warned you not to do certain things. He did the precise thing which I warned him not to do. I hope you will not follow his example. It was a very natural mistake that Ben made. He said to himself, as he told me the other day when I spent a pleasant evening with him, that the way he looked at it was this: that the prairie lands of this great state would never be settled up. He told me that just as he left the East he dreamed that, in moving West, when he came to the top of a hill, the left back wheel of his wagon came off, and that before him there lay a beautiful stream with timber growing along its banks, a log cabin and a fertile prairie for miles on each side. The dream was fulfilled when he actually began his journey, and there lay the landscape he had seen in his vision; and when he had repaired his wagon he called on the owner of the cabin and bought that quarter.

He was one hundred and fifty miles from a railroad. He said to himself: "Here is timber, shelter, good land, and pasture that would have made Jacob weep if Esau had squatted on it. What more do I want?" In less than five years from that time the railroad came, and with it the settlers jumping over each other to enter the land, and he was shut up to his timberland, which he has been grubbing out every since. "I do not mind," said he, "the hard work it has cost me

and my boys to prepare the fields, when, had I gone out a few miles, I could have had much better lands at government price, in which I could have plowed the length and breadth of a quarter without striking a stone or a stump; but there are a lot of folks settled along this timber, with some of whom I do not care to have my boys and girls associate, and I must get out of this as soon as possible if I am to have a happy and peaceful old age."

His brother Sam went out on the prairie and Ben furnished him firewood free for five years. He built a temporary house and prairie stable, broke up his land by piecemeal, and finally got it all under cultivation, and along early in the sixties had the misfortune to have a great wheat crop and sell it at a long price. This awakened his dominate ambition. He thought he must have more land, and bought an adjacent quarter, giving a mortgage on both. The next year the wheat crop failed and the price went down. Bob Cheatem, the son of a broker of the firm Ketchem and Cheatem, whose acquaintance Sam made in coming West, and who had a large flock of diseased sheep on his hands taken in on a mortgage, called to revive old acquaintance with him, and incidentally persuaded him to go into sheep. Sam knew nothing about sheep, but yielded to Bob's persuasive eloquence, which he describes as follows: "You see, Sam, wool is worth a dollar a pound, and every ewe will shear eight pounds each year, and give you two good lambs. The lambs are worth five dollars apiece, and there is eighteen dollars for keeping one sheep a year, and you can keep six of them on an acre." Sam bought the sheep and millions of scab mites with them, and foot rot to boot; and in less than a year he sold all that remained of the flock for a dollar a head, and was glad to get rid of them at that. Bob had a second mortgage on the farm, and a chattel mortgage on everything not exempt from execution. Poor Sam has been working from that day to this, year in and year out, to get rid of that mortgage given for Bob's sheep.

The trouble with Sam was that life had been too easy with him in boyhood, and a little prosperity made him dizzy, as it has made many another man. He had never really studied farming, and when misfortune came, he grasped, like a drowning man, at a straw, was easily the dupe of a designing scoundrel, and went into a department of farming of which he had no knowledge whatever.

I count his boys more fortunate than he. They are experiencing misfortune when they are young; but if they have grit enough not to be disgusted with farming, and sense enough to look around and see that other farmers prosper who follow out the right lines, they may one day make the name of Hardup a misnomer in their case.

Their brother Jim was a fortunate fellow. He married a bright, snappy little wife, with eyes that blaze like coals or fire, or make a fellow's heart go pitty-pat when she looked at him lovingly (I used to see her home from singing school occasionally), and she took Jim in hand, and made him, as they say, "toe the mark." I knew she would do that. There was no sleeping until after sunrise in that house. Jim had worked and she managed, and it is a joke in the neighborhood that when a man wants a little money he is directed to go to Jim Hardup. Whatever mistakes the rest made, Jim made none when he married. Had he married one of the clinging ivy sort that say, "Not as I will, but as I please," Jim would have been as hard up as any of the Hardups, and his boys would have lacked the grit and snap that make their name a misnomer and a standing joke. Ben's boys and Sam's are renters; some of them hired hands. Jim's boys own their own farms. Moral: if you are ever inclined to be easygoing, do not be afraid to cultivate the acquaintance of the girl that has more get up and snap than you have; it may be the life of you, young man, and I think it will.

The Hardups, however, do not all live in the country. I know plenty of them in town. Some are chronically hard up because they have made mistakes in the past and cannot help it. I pity them, as I do every man that has been unfortunate, through his own fault or not. Some are financially hard up for the time being because of sickness or other misfortunes. Some because they have been too honest and scorned to take an advantage that would have made them rich. Others because they have had too much faith in human nature and have been victims of scoundrels like Bob Cheatem, who live by studying the weak points of their fellow beings, winning their confidence, and robbing them under the forms of law. Many a fine house stands on a corner lot on a fashionable street in the city, built with money which was never earned by the owner, but stolen under forms of law from the men who earned it; and these men now, if properly named, would be called "Hardup."

I shall never forget the look that came on Sam Hardup's face, when, in passing through his country seat we came to Bob Cheatem's broker office, miscalled "bank," and saw his fine horses and carriage, with the liveried coachman, standing by the curb. He stopped and pointing with quivering finger said, "There is the scoundrel that has made me and mine poor. May his wife be a widow, and his children fatherless." I was about to rebuke him when it occurred to me that he was quoting from one of David's psalms. As we passed on he said: "Henry, you will have to excuse me this time,

but nothing but the so-called cursing psalms meet the man's case; and I think it was to describe just such scoundrels as he that they were written. He owned those scabby sheep, and in pretending to give good advice to a friend in trouble, made him poor for life. 'Let his iniquity return upon his own head.'"

And I said, "Amen."

Uncle Henry

My Dear Boy:

I told you in my last letter about some of the misfortunes that befell various typical members of the Hardup family. Whether they are true to type or not, you can very easily find out by observing various members of that interesting family within range of even your limited acquaintance. You may possibly be interested in some mistakes that have been made by various typical members of another family equally ancient and honorable—the Richmans. The Richmans are not nearly so numerous as the Hardups. For some reason there are comparatively few of them. Abe Lincoln used to say that he thought the Lord must like the Hardups best, or he would not have made so many of them. For some reason the Richmans have, usually, small families; and the more exclusive and aristocratic they become, the fewer children they seem to have.

They are a very old family. We read in Bible times of Solomon Richman. If he had not plenty of money, I suppose he would have gone by the name "Sol." He knew a lot more than any man of the family that I ever heard of. He was regarded as the wisest man of that day; and yet in his old days, in looking back over his past, he seemed to put very little store on his money, saying in effect, that the piling up of money was vanity and vexation of the spirit, for the reason that no man could tell whether his boy would be a wise man or a fool, or words to that effect; that the man who gave himself up solely to piling up money never knew who was going to spend it, and that the very best thing a boy or man could do was to fear God and keep his commandments. In this opinion, I concur. He did a great many foolish things, but, taking all in all, I regard him as the smartest specimen of the Richman family I ever heard of, and I advise you to read his book on the conduct of life, which is generally known by the name Proverbs.

I was reading only last night of a member of the family who died a few days ago in England. He began life poor, was the son of a coal miner, and had scarcely enough to live on the first year of his life. He went to school at night, lost his father in early youth, but became one of the greatest men England ever produced, dying at the ripe old age of ninety-two, and as cheerful in his last days as a boy of twenty.

While this family numbers some of the very best people the world has ever known, it numbers a lot of very great scoundrels. There seems to be something wrong with the breed. They are not like the Hardups, an even lot; and it is somewhat notorious that their boys seldom turn out as well as those of the Hardups, and their girls are very liable to make poor matches. I suppose this is why they are such an uneven lot.

One of my earliest friend was Colonel Alexander Richman. He was a farmer, and got his title, not by service in the army, but as colonel of the militia. He was a very good farmer, indeed, one of the best I have ever known, and being a good businessman as well as a farmer, reading the agricultural papers of that day very closely, watching the markets, and keeping his credit away above par, he made a lot of money. He branched out into matters outside the farm, and for many years made money hand over fist without oppressing anybody, or taking mean advantage. His word was as good as a government bond. When his boys got hold of the business, backed as they were by their father's unlimited credit, it spread out, so to speak, all over creation, with the result that in a few years the entire credit of the family was no greater than that of the poorest Hardup in the neighborhood, and their cash not much greater in amount.

I knew his brother John well. He was a well-to-do farmer, had two hundred acres of land of the best, with the finest improvements, a fine brick house, large and commodious barns, a great orchard, every field fenced hog-tight, and everything else to match. He had an only son named Robert. His mother died when he was a baby, and he was brought up by two maiden aunts, in whose eyes nothing was too good for Robert. He slept late in the morning, and had an elegant pony to ride, fine clothes, and all that. None of the neighbors' girls were good enough for him, and he married a lady of reputed wealth a long way from home, who knew nothing of farm life, and he had to keep two girls to wait on her. When the new wife came the aunts paid dearly for their indulgence of Robert, and left, calling him an ungrateful wretch. In a few years a young Hardup, who was getting on in the world, took in by sheriff's deed the last

forty of Robert's magnificent inheritance. Robert moved to town, and died a wreck.

A distant relative of this same family moved West. His mother, so the tradition among the old folks goes, had a fancy for odd names, and she called him Graybel. He was known among the boy as "Grabe" and rather well-liked. He had none of the aristocratic airs that characterized the other members of the family. In fact, he became quite popular in school. His father was one of the best of the connection, and, in trying to uphold their credit with his own, failed, and young Graybel moved West, starting in the world poor. He worked late and early, never went in debt, lived poorly, and married a thoroughly good, quiet sort of wife, of whom he, as well as his children, subsequently made a slave. When he got a little money ahead through working, scraping and saving, he loaned it to his neighbors at anywhere from two to five percent a month, a thing not unusual in those early days, and took cutthroat chattel mortgages, iron-bound, copper-bottomed, and warranted to hold anything except a man's life in his body. As his wealth increased, he loaned to his neighbors who owned real estate but were a little behind-hand on the same sort of ironclad mortgages, making all payments due and payable on default of the principal of interest of the first payment, and foreclosed on the first opportunity. He became wealthy rapidly. The love of money took complete possession of his entire being. The demon of avarice took an ironclad mortgage on the entire family, except his patient and long-suffering wife, who was charitable to the extent to which she could carry on her benefactions in secret—a limited extent in that family.

Father and sons worked together with one mind and purpose, drove hard bargains, bought stock, land and grain at the very lowest prices which their owners were compelled by their hard necessities to accept, prying constantly into the business of their neighbors to see how soon, and to what extent, they could put on the screws. On weekdays they wore the coarsest clothes, and it was often remarked on the quiet that but one of the boys was ever seen at church at a time, the old man never, and that the same suit of clothes seemed to fit all the men of that family. Finally the wife and mother died from sheer overwork and exposure. Her last remark was: "I am so tired—so t-i-r-ed."

In less than a year a second wife appeared on the scene, but she did not linger long. She was smart, ambitious, fairly well-educated, liked to dress in good taste but not extravagantly, had a temper of her own, and a tongue that could cut like a razor without even raising the tone of her voice. Long before this happened the neighbors had changed the name of Graybel to Grab-All. There was some quiet talk in the neighborhood when Grab-All Richman had his hair dyed, put down new carpets, got a good suit of clothes, and a fine new buggy; but before the hair dye had disappeared, and long before the carpets or buggy were worn out, there was a first-class sensation in the circuit court in the shape of a divorce suit, in which Grab-All Richman was defendant and his wife plaintiff, and a decree for alimony which required the sale of two good farms to satisfy. This broke the old man's heart, and he died in the winter. Before the grass grew in the spring on the sod which covered his grave, the sons were at the swords' points over the division of the estate, and there was a public washing of soiled linen that disgusted the entire neighborhood.

You will not live many years, nor become acquainted in very many neighborhoods until you find families and individuals that approximate to this type of the Richman family, though I hope you will not meet any which this description entirely fits. I draw the picture that you may learn how to shape your life so that its ending will not have the faintest likeness to that which I have drawn.

The Richmans, however, do not all live in the country. Very few of them, in fact, do, the atmosphere of the city being much more congenial to their aristocratic tastes, and city conditions much more favorable to the gratification of their chief ambition of most of them, that of making money, or rather of transferring money from the pockets of other people to their own. I was walking along one of the fashionable streets of one of the largest American cities recently, with my friend, Silas Richman, who, by the way, is a bit of a philosopher. He called my attention to a number of his relatives and connections, who were driving along in the fashionable boulevard with their fine teams driven by liveried coachmen, and said: "Please note the lines of care and anxiety written on the faces of those men, and contrast them with the happy-go-lucky air of the people who are walking on this street. I have studied this matter closely for a number of years, and have always found that the fellows who bear heavy burdens ride in that style, while the men with light hearts, happy countenances, and free from care, and who have real enjoyment in life, walk. I used to train in that crowd. I made in twenty years over three-quarters of a million dollars. I have lost $725,000 of it, and my health besides, and I am just beginning to realize what a consummate fool I was, and what consummate fools are the relations of mine. In 1893, I had $830,000, good value. I knew I did not need anymore, but I took the foolish notion into my head that I must be able to truthfully call myself a millionaire. I made large investments and involved myself in debt in order to make that other $170,000 at one bold, Napoleonic stroke. I had paid insurance for twenty-five years, and never lost a cent; but just at the wrong time one building took fire on which I had foolishly allowed the insurance to lapse two weeks before, and it swept away $130,000. In a month another came and swept away $64,000 more. This shook my credit, and I was obliged to sell property at a sacrifice. I had built a residence costing me $90,000; spent $21,000 in furnishing it; had fine teams and carriages, and was starting out in great style. When one piece of misfortune after another came, I began to realize my folly, and figured that I could board myself and family in comfort for the taxes and interest I was paying on my establishment. I want to say to you now that the last year, when I have been living sensibly as a common sort of man, has been the happiest year of my life.

"I was no greater fool than the rest of them are yet. Look, for instance, at my cousin George. He was reputed worth $40,000,000. He died suddenly last week, a comparatively young man. When he is "cut up," that is, his estate divided, it will probably be found to be less than $10,000,000. His sons are drunkards, and unless he has them cut off, which I suspect he has, with a life annuity, his property will go to the dogs.

"His brother Charles is reputed worth twenty million, and is probably worth eight; has been once in the penitentiary, and twice bankrupt. Each of these men has incidentally rendered great service to the public while getting rich, but they are hated and despised because of their avarice and greed. Charles told me only last night that if he was as bad a man as people thought he was he would drown himself before morning. The fact is that fate, perhaps you would call it providence, makes out of their avarice, greed, and ambition, whips and goad to compel the Richman to carry out great enterprises of which the public receives the benefit, and for which they get only curses. As for me, I am gathering up what is left of the wreck of my fortune, the result of my foolish ambition, and I propose to take what comfort I can in this world while I am left in it, and regain my health if I can."

I would not have you believe that all the Richmans are of the types that I have sketched. Many members of this family are among the best people that I have ever known. The trouble with them is that they do not run evenly, and have not a clear, well-defined moral type. If they were all good men this world would be a great deal better world than it is.

I like to see men make money and plenty of it, honestly. I like to see farmers, merchants, and all sorts and conditions of men who follow honest calling and use honest means get on in the world. Capital is essential to the proper conduct of the world's business and is one of the best friends the poor man has when handled by honest men. I hope you will be a rich man some day, even if your name is not Richman; but I could not wish you a worse fate than will befall you if you set before yourself money, profit, wealth, as the end to be desired above all other things, and at the expense of honor, manliness, and character.

When this passion for getting money in any way possible, but *getting it* takes hold of a boy or man, it is sure death by strangulation to every noble purpose, and every instinct, even, that distinguishes man from the swine he feeds. It renders him false to his associates—true friends he can have none—cruel to his family and to his hired hand. He must of necessity be a harsh and cruel husband; a father whom his sons and daughters may fear, but can never love as children should love their parents. Those whom he has wronged hate him; those who know him best necessarily despise him; and his memory, like that of the wicked, shall rot.

Uncle Henry

My Dear Boy:

When I was a farm boy we had in our neighborhood a representative of the Hardman family. I suppose in my innocence that this was about all of that family or class there were in existence; but I have since learned, and so will you, that they are a large family, very widely scattered all over the world, and quite ancient, if not quite honorable. Even as a boy I noticed that one of the peculiarities of this family was that no one really liked them, or even pretended to like them, unless he had something to gain by it, or fear from them, if he failed to pretend to like them.

I could never discover that old Jakie (no one never called him Mister) Hardman, had a true friend. I could never discern that his boys had any affection for him, and, judging from the way his daughters ran away with worthless scamps, I took it that fear and not love ruled in that home.

After I left home and met with other members of the Hardman tribe, I discovered several other peculiarities. One was in their

choice of lines of business or profession. I never knew one of them
to become a doctor, preacher, or professor in a college, and seldom
one a schoolteacher, and he for not more than one or two terms. I
have known a few of them to become lawyers, but always in connection with some other line of business such as note-shaving or
real estate. I never knew as much as one of them to be an editor, but
one or two were business managers of newspapers—for a time. I
never knew one of them to be a candidate for Congress, or for the
state legislature. I have known a number that were members of city
councils, and quite a number who were assessors in the city, but
never one in the country.

Another peculiar feature of the Hardmans is that they do not
generally believe in any education beyond the "three R's," readin',
'ritin' and 'rithmetic. I have known, among the hundreds that I have
met, a few that were fairly good farmers, none that were fairly up-
to-date; but as a rule they got rich and made more money than the
up-to-date farmers, not, however, by farming, but by trading, and
by loaning money at the very highest rate of interest, and on cutthroat mortgages.

Perhaps the best way to give you an insight into the Hardman
character is to tell you the story of Tom Hardman. He was not a
bad boy, as I remember him, and had the sympathy of most of us,
because we knew he had a hard time of it at home. He was worked
hard; driven like a slave, in fact, from the time school closed in
March until the beginning of the winter term in December. He was,
however, naturally bright, and picked up knowledge quite readily.
He was sharp trader even then, and the boy who swapped knives or
caps with Tom Hardman always got the worst of it. The worst thing
I knew about him in those days was his disposition to knuckle to
the big boys when he ought to have resented their insults, and his
tyranny over the small boys who had no big brothers to take their
part, two things, which, in my observation, always go together in
boy and man. He ran away from home at the age of fifteen, and
after I made the acquaintance of large numbers of the Hardman
family, I was all the more anxious to learn Tom's history.

On my last trip abroad I noticed on the passenger list the name
of Thomas Hardman, Esq., and was glad to learn, on introducing
myself, that he was none other than my old schoolmate. An ocean
steamer furnishes one of the best opportunities to study human
nature and find out what men really think on all important subjects.
Passengers are completely cut off from the outside world for a week
or ten days. There are no letters, telegrams, no press of business,
nothing whatever to do but to be seasick, eat, sleep, look out for

whales or icebergs, and tell stories to kill time. There is more or less of an element of danger to all, which draws people close together and makes them willing to reveal their true character. Neither Tom nor I were seasick, and after we had traced out the history of each of the old boys and girls in detail, we began to unfold our own experiences. He was not free to talk about himself at first. I felt my way gradually by talking about matters of current history, such as the probable working out of the "Wilson Bill" then going into effect; then on partisan politics, literature, manufactures, and finally on agriculture. I told him of my own hopes and ambitions in the line of newspaper work; that my aim was to develop the agriculture of the nation, and especially of the West; to aid in developing a class of farmers mightier than Caesar's legions, more invincible than Cromwell's Ironsides, the stay of the country in war, its balance wheel in peace when other classes lose their heads; and that I wished so to live and work that when I was dead and gone my name would be remembered by thousands as a man who had left the world better than he found it.

He finally said to me on the last day of the voyage: "Henry, you have been a fool all your days. You had it in you to make money and plenty of it; but you have chosen instead to run a fool's errand by trying to help other people. I have heard of your doings from time to time. I know more about you than you think, and what I say to your face I have said dozens of times behind your back. You are a fool. I have no doubt you have helped many, or at least you think you have. You have also loaned money on poor security, and you have been too chicken-hearted to put the screws down hard and realize on what security you had. You have let women cry you out of forcing collections, and they have laughed at you behind your back. What do these people care for you or yours? You have helped men into place and power, and they have kicked you; you have given scoundrels your confidence, and they have betrayed it, and slandered and abused you in order to make themselves believe they owed you nothing."

"I have done nothing of the kind. You and I are as wide apart as poles. You believe in a God; I do not. You believe there is a future; I do not. You believe there is a right and wrong; I do not. You believe there is such a thing as sin; I do not. If there is a sin, it is that of perpetuating the race in such a cursed world as this. You started out to look after other people and to teach them how to fit themselves for another world; I started out to look after number one, which means *me, myself,* and I have *done it.*"

"I ran away from home, as you know. Why I did it you know. My father never loved me, and I never loved him. There was no love between my father and mother, brothers and sisters. Love of every kind is a fool's dream; modesty is hypocrisy; humility, cowardice. I hired out to a farmer out West at $200 a year and board. I drew enough to live on, never over $50 a year, and often much less, and I took his note to ten percent per annum for the rest, and then loaned him the interest. I did this for ten years. The tenth year he had hard luck. His crops failed, his hogs died of cholera, and his cows aborted. Times were hard, neither the banks nor loan companies were advancing money, and I foreclosed and took the farm subject to a mortgage of $2000 which cut out the homestead rights and his wife's dower. I farmed and he hired out."

"My credit was now established. I could borrow when, and as much as, I wanted to. I quit work, rented the farm for cash rent with an ironclad lease, and collected every cent, although it took all the fool fellow had and left the judgment still unsatisfied, which made him slave for a year or two more. I looked out for lame ducks and took them in, and made money hand over fist."

"I soon got tired of skinning grangers. They squeal when they are skinned, and so do the neighbors. They have a lot of old fogy notions in the country. They think the Ten Commandments are binding, and that Christ talked business in his Sermon on the Mount. At thirty-five I was worth $20,000. I shook the dust of the country from my feet and came to the city. I came in on the Wabash. The train was detained by wrecks—a freight wreck in front, one behind, and one on a branch line, and we had to wait half a day. I fell in with the master mechanic of the shops in the city, and we got to talking about Jay Gould's management of his railroads. He told me that Gould managed to borrow, or get proxies for enough of stock to elect him president, and he then in various ways decreased the revenues by large salaries, by improvements, by diverting traffic to other roads, until he run the stock down, and bought in as much as possible. Shrewd men were afraid to own stock in anything that Jay Gould controlled, and he took advantage of his own rascality. When he got a large amount of it in, he began building up the road by reversing his methods, and sold out the suckers. He thus milked the public into his own bucket. I said to myself, Why cannot I do this on a small scale? The first thing I did was to find a corporation lawyer after my own heart, and agree to give them a slice on the sly. He was well-acquainted in the city, and we looked out for prosperous corporations and found one who would sell enough stock to give me a controlling interest by voting with one faction or the other. We then made our deal for directors with the side that was the more easily deceived. When we got control we put up salaries, wiped out the surplus, made no dividends, and rendered the stock worthless to the minority. If suits were brought, my legal friend, whom I employed to tell me, not what the law was, but how I could evade it, demurred, delayed, postponed and worried the other side until they sold me their stock for a song. I soon found that I could swipe in a hundred dollars in the city quicker than I could ten in the country. The best thing of all is that city men do not squeal. Their code of ethics is the commercial, not the moral. Their motto is 'dog eat dog,' and hence in the city dogs are respectable. If a man attends a fashionable church, and is good pay, he can do about as he pleases. If the preacher has old fogy notions, and talks old-fashioned morality, he soon gets a sore throat, or his wife needs a change of climate. It is not so in the country. The stupid granger looked on my proffered donations as 'the price of a dog,' or 'the hire of a prostitute,' quoted scripture, and said Tom Hardman was trying to buy his way into heaven with the wages of unrighteousness. The city is the place for me. There is ten dollars to be had by looting corporations to one dollar by skinning grangers."

"Do you mean to tell me," said I, "that there is no relief in a corporation for minority stockholders? I myself am a minority stockholder in a newspaper corporation, and there is trouble ahead with the majority."

"None, whatever, unless you can prove the most glaring fraud. The majority can defraud all they please if they have the right kind of lawyer. He can file motions for more explicit statements; can move to strike out part of the pleadings, or divide, and thus secure delay; he can then demure, postpone, appeal, ask for new trials, prolong litigation for ten years, until the property is entirely eaten up in salaries and expenses. The danger of being caught in fraud in a corporation never troubles me. I never give it a moment's thought."

I looked at him in amazement and replied, "Tom, this is the first time we have met for forty years. It will, in all probability, be the last. I will not put in words what I think of you and your methods. They are not new. They are as old as the Egyptian bondage. They are the methods of scoundrels in all ages. They have ruined not men merely, but nations and civilizations. The prophet Micah described just such scoundrels as you when he wrote:

Woe to them that devise iniquity—and work evil upon their beds:
When the morning is light, they practice it,
Because it is in the power of their hand.
And they covet fields, and take them by violence;
And houses, and take them away:
So they oppress a man and his house—even a man and his
heritage."

"He described you to a dot when he said:
Who hate the good, and love the evil;
Who pluck off their skin from off them—and their flesh
from off their bones;
Who also eat the flesh of my people—and flay their skin
from off them;
And they break their bones—and chop them in pieces,
as for the pot."

"It is men like you that are corrupting the very foundations of public morality, and fast bringing about the same condition of things which the prophet described when he said:
The princes abhor judgment, and pervert all equity.
They build up Zion with blood—Jerusalem with iniquity.
The heads there judge for reward—and the priests
thereof teach for hire,
And the prophets thereof divine for money:"

"You are a typical Hardman, not only in name and nature, but altogether the meanest, lowest, and most dangerous that I ever met; and notwithstanding all your disbelief in a future or a God, or the worth of this present life, if this ship were to strike a rock and begin to sink, you would be the first man to push the women and children out of the lifeboat in order to save your worthless carcass."

You will, my dear boy, in the future meet with many of this family. You will not find many as totally destitute of all human feeling and sense of honor as this one. I wish, however, to put you on your guard against the whole tribe, for when a man hardens himself against right, and uses his full power to oppress, whether his motive be the love of money, or the love of power, it is only a question of time, opportunity, and ability when he will fill out the full measure of the iniquity of Tom Hardman. It is only a question of time when the man who abandons the moral code which has made this nation great, and gives himself over to the teachings of commercial morality, current to some extent in the country, and more largely in the city, will become a moral wreck, deserving of the scorn and contempt of all men who love their country or their race.

Uncle Henry

My Dear Boy:

In my last letter I quoted Tom Hardman as believing in what he called "commercial morality." This may possibly be a new term to you and need explanation. You have probably assumed that there is but one kind of morality, that which is taught in the Sabbath school and from the pulpit, is based on the teachings of Moses and the prophets, and finds its best statement and application in the Sermon on the Mount. Before you have any very large experience in the world, you will discover that there is another morality, practiced but not preached, that pervades very largely the business of the nation, of our own nation, and particularly of the large cities, to no little extent of the country, and, to some extent, of your own particular neighborhood. No pulpit proclaims it, no Sabbath-school teacher mentions it, no newspaper advocates it, no individual avows it until he has reached a point when he feels it safe to defy public opinion. With this exception, the only men who are the avowed believers in this commercial morality are common thieves, confidence men, gamblers in common gambling houses, gamblers on the boards of trade, and such other professions as are under the ban of public opinion. The most common hypocrisy practiced in these modern days is that of professing to believe in Christian morality, and yet in business practicing commercial morality, and making atonement or compensation by liberal contributions, obtained by practice of the opposite, in support of the teachers of Christian morality.

To make this plain, let me say that Christianity assumes that there is a possible right and wrong in every business transaction; that the moral law governs there, as elsewhere; that there can be no separation of business from morals, and that a fair and just trade is as pleasing to the Almighty as a church contribution, a prayer, or a sermon. If the Bible does not teach this in its every page, directly or indirectly, I confess I have never been able to understand its meaning. Commercial morality, on the other hand, assumes that business is one thing, benevolence another; or, to put it in the tersest possible terms, business is business; by which bad men mean that business has no connection whatever with morals or religion. Good men frequently use the term with an entirely different meaning, namely, that business should be conducted on well-established business principles. Now the truth is that while business should be conducted on the principles which human experience for ages has proved to be correct, nonetheless will a business conducted on these principles prove to be one of the highest forms of benevolence, in that it will encourage thrift, self-control, integrity, and furnish reliable and steady employment to the thousands that are not capable of conducting, on their own account, large business enterprises. While business is not religion, nevertheless it furnished the best possible sphere for the practice of the basic principles of all religion, taught in every pulpit in Christian lands. If we divorce business from religion, we cut the very foundation from under all civilization that the world has yet achieved that is worth retaining.

The danger to the farm boy is that he may adopt the maxims which I have quoted above in their bad sense, instead of in their true and proper sense; and it is only a question of time, opportunity and circumstances, when, if, he should do so, he will develop a character by which Tom Hardman is extreme, but by no means an uncommon type. The foundation of all dishonest business is in buying a thing for less than it is worth. Every thoroughly honest trade gives a full equivalent for the value received. I wish you to get this idea clearly in mind. Honest dealing consists in buying things for what they are worth, as determined by the supply and demand; dishonest dealing consists in getting in some way the advantage, and buying for less than they are worth. Every trade on its face purports to be the exchange of goods, or goods for money, as of equal value, or as a full equivalent. You say: How is it possible for me to make money or profit by trading, if in every honest deal a full equivalent is given? The answer is easy and can best be given way by illustration. Your father raises corn. He grows more than the family, or the stock on the farm, can consume, or than he desires to keep for an advanced price. He sells this corn at its market value on the date of sale as determined by the supply and demand in great markets. He can not use or keep it to great advantage. He therefore sells it to the man who has use for it, either to feed to his stock, or to ship to a distant market to his stock, or to ship to a distant market at carload rates, which are always less than rates on the part of a carload. While a full equivalent is rendered, your father is the gainer, because he has disposed of something for which he had no present use. The money that he receives for it is of more value to him than the corn, while the corn is of more value to the buyer than the money; hence, both profit by this strictly honest trade.

Your father grows livestock as a means of disposing to better advantage the products of his farm. When it is finished he can to use it to advantage. It has gained all that it can profitably. He ships it to the nearest stock market and sells it to the packer. Your father can use the money to much better advantage than he could the livestock; hence, each profits by the transaction, for each has disposed of an article for which he has no present use—your father, the livestock, the packer, the money.

Your father wishes to buy the livestock to consume his grain, but does not have the money; the banker has. He gives his note for six months at six or eight percent interest, figuring that after paying the interest, and paying himself market price for the corn, and a reasonable price for the grass that it will take to fatten the stock, he will have more left than will pay the interest on the note. The banker puts the money to work for him in the way of bringing in interest; your father puts the money to work in condensing his crops for market; hence, each is a gainer by the transaction, and neither would enter into it unless he had reason to believe he would be the gainer. Each has rendered to the other a full equivalent, and by reason of their different circumstances and conditions, each makes a profit. The same law applies in all kinds of legitimate business. A full equivalent is rendered in every case with the possible and probable profit to both parties in the transaction. All such transactions are in accordance with Christian morality.

Men who are guided by commercial morality act from entirely different methods. The idea of getting either something for nothing, or much for little, is the prevailing motive with them. For example, the manufacturers of any line of goods form a trust. They close up factories, dismiss labor, limit production, and advance the price, the object being to secure from every customer something more than the article is really worth, or for which it can be produced. They think they *can* do it, and proceed to do it without the slightest regard to the rights of labor, or to the real cost of production, and on the theory, which is the essence of savagery and barbarism, that *might* makes *right*. The robber barons of the Middle Ages who occupied commanding positions on the great highways of travel levied blackmail on all comers and goers *because they could*. Their legitimate successor are the robber trusts of the nineteenth century, who take a few cents from this man, and a few dollars from that, simply because they *can*, or *think* they can. This is commercial morality.

The railroads have a large amount of what is known as watered stock; that is, certificates of stock issued without the value of the face of the stock being expended in constructing the road. In other words, it was issued without consideration. In order to secure dividends on this watered stock, they have been forming combinations for the last twenty or thirty years, agreeing to advance and maintain rates and compel the public to pay the increase. The only justification made for this is that they *can*. We find, when we get down to the very truth, that the basis of all modern rate-making is what "what the traffic will bear;" that is, what the public can be forced to pay. This is commercial morality—the morality of the robber baron of the Middle Ages, the morality of the thief and the robber. I *can*, therefore I will. Or, as Rob Roy puts it: Let him take who *will*, and keep who *can*.

Corporations of all kinds take kindly to commercial morality. A corporation is an artificial person. It is made up of stockholders who own shares, and it exempts the shareholder from any personal liability beyond the value of his shares. Its life is limited to the years prescribed in the charter by the law, but provides for a renewal indefinitely, and a majority vote of the shareholders governs. It is thus endowed with practical immorality. Death, that cuts short the robberies of the individual, spares the corporation. It has no soul to be saved, or to be lost, and hence is very likely to ignore all moral precepts, all idea of responsibility to a higher power, and very gradually develops among men who have much to do with corporations, what is known as a "corporate conscience"—a conscience that has no regard for moral law, and but little regard for human law. For it a curious fact that a man who repudiates moral obligations has little respect for legal obligations. The more experienced he becomes the less respect he has, owing to the fact that he finds out how easy it is to evade legal penalties by methods which I have in a previous letter described. Many men, in fact, have two consciences—a corporate conscience, and an individual. As members of a corporation, they will do things, apparently with a clear conscience, which they would absolutely scorn to do in the transaction of their private business. In the one case commercial morality, or the morality of the thief and the robber, governs; in their private business Christian morality governs until the greater immediate gain to be made by corporation methods blunts the Christian conscience, and they become business hypocrites, professing one thing, and practicing the opposite. Hence, in large cities business is becoming largely "dog eat dog," the men in one line of business waiting patiently until sharpness of competition results in the failure of a competitor, and they all pounce on the crippled man and devour his substance, much as a pack of wolves stop in their chase to devour one that has been shot or pursued.

I was sitting, one evening, on a deck of a steamer on the Pacific, as the cooks were clearing off the tables and throwing the scraps out of the porthole into the ocean. Dozens of white gulls were following in the wake of the ship, and dropping down and devouring bucketfuls of scraps as they were thrown on the waves. One large, gray gull, of an entirely different species, followed along leisurely, and just as the white gulls began to devour the coveted morsels, dropped down amongst them and scooped everything into his capacious crop. He kept this up for an hour, and I marveled at that bird's capacity, and said to myself: "There is a type of business life; that scoundrel waits until the gulls have located the food and had a taste, when he swoops down and takes in the bulk of it." By sheer force of power he was the robber of the sea, and a fine illustration of a class of men who are governed only by commercial morality.

Nor is the country exempt from the operations of men guided only by commercial morality. The horse jockey is perhaps, the best type—a type, however, so well-known by honest men, that he is never trusted by farmers. They never believe what a horse jockey tells them. They, perhaps, have never stated the reason to themselves, but it is because he is guided by commercial morality. They do not believe him even when he is not trading horses. They are right in this, for a man who learns to deceive in one line will soon learn to deceive in all. A more responsible type, but more dangerous, is the broker who calls himself a banker and extorts usury, anywhere from one to five percent a month, because he *can*. When I hear farmers say that such a broker, or so-called banker, is not "in business for his health," I know exactly what they mean. A still more dangerous type is that of the respectable farmer who practices the broker's methods. He has money to lend, not to the best farmers, but to the worst, whenever their hard necessities compel them to extortionate interest. This scoundrel often creates necessities by urging men to borrow when he knows that borrowing must lead to less, extending credit to an unreasonable limit, and when he finds the borrower in a tight place by reason of crop failures, or other misfortunes, puts on the screws and demands immediate payment; and in default, forecloses, bankrupting the borrower. Another example is that of the landlord who uses all his power given him by a landlord's lien, either to bankrupt the renter completely, or to hold a judgment over him in such a way as to make him his slave for years to come, and all because the law gives him the *power*, which he mistakes for the *right*.

I have thus far spoken of men who practice commercial morality with a set purpose for gain. I would not do justice to a large class of businessmen were I to fail to state that many of them are compelled, in a manner, to practice commercial morality or go out of business. Let me illustrate: A large stored in the city, for example, that repudiates Moses and the Sermon the Mount, and believes thoroughly, in a bad sense, that "business is business" offers for sale, at a low price, goods that have been dishonestly made by contract, possibly in sweat shops, or in factories where shoddy displaces honest material and where workmanship is cheap and poor. This class of stores sets the pace which honest men must follow, or go out of business, at least until commercial morality is so far educated out of buyers that they lose their mania for buying bargains. Until this is done, the dishonest element in business will set the pace which honest men must follow, or quit. These dishonest dealers compete with each other in the race of securing cheaper and more worthless goods, by cheapening material, lowering the price of labor, forcing honest, but poor, laborers into pauperism, and honest and skilled laborers to accept the wages of the unskilled, thus degrading labor, demoralizing business, debauching the public morals, and transforming us into a nation of adulterators, money-grabbers, bargain-seekers, and all that, until the problem of how to be an honest merchant, and practice the Sermon on the Mount on weekdays while professing it on Sabbath, is one of the most difficult problems of human experience. I seldom hear a lady boasting of how cheap she bought a dress or bonnet, without thinking of the poorly paid woman who made that bonnet. Lazarus must work cheap, beg, or starve in order that Dives may fare sumptuously every day. The trouble is that the mania for cheapness, the craze for the bargain counter, pervades the city and country alike; and when we come to the last analysis it is closely related to the gambler's mania of getting something for nothing. I did not intend to moralize in this way, but it is better that you should understand before entering on active work some of the difficulties and perplexities with which the businessman must grapple if he intends to be a thoroughly honest man.

It is a very healthy sign that in nearly all country communities men who follow these practices are more or less under the ban of public opinion, an opinion not always expressed, but felt. One of the highest compliments that farming communities pay to themselves, is the high honor in which they hold farmers and businessmen of all classes who do business on principles of highest honor. When a man sends a carload of hogs or cattle to the dealer at the station in the full confidence that whether the market of the day before be up or down he will get the full value without a previous contract, he pays him about as high honor as one man can well pay another, and we have noticed that dealers who treat farmers in this spirit are uniformly men who make money. In all dealings of man with man, the confidence of the customer is the most valuable asset of the dealer. It is something that cannot be taxed, or destroyed by fire, or by flood; it cannot be measured by dollars, but is gradually coined into dollars as we transform the rain, the sunshine, the electric currents, and the fertility of the soil into crops. There can be no confidence, whatever, reposed in the man or corporation that is guided only by commercial morality. It is death to manhood, death to legitimate business, death to every noble feeling and aspiration, and were it generally practiced, or even nearly so, it would be death to the civilization of the nineteenth century. It is under the condemnation of every law of God; it is under the ban of all good men; it is civilized savagery and business barbarism; at least so believes your Uncle Henry.

My Dear Boy:

When I was your age, although I had been very well instructed in the doctrine of "total depravity" as a theological proposition, I did not believe that such men as Bob Cheatem and Tom Hardman had any real existence in country places. I thought they belonged to the city. I thought the doctrine of "total depravity" as a theological proposition had to be modified in a great many ways to make it conform to the facts of existence. I had never heard of "commercial morality," but a great deal of the morality taught from the pulpit and in the home. I was taught that while, as an abstract principle, men were totally depraved, and sinned as soon as they were born, if not before, or at least, as soon as they were able, nevertheless, it was but just and fair that men should prove themselves bad before I had a right to treat them as bad men. It has cost me a good deal of money and grief to learn that I was mistaken in some things, and to discover that, even amidst country surroundings, among farmers and farm boys, types of the characters I have mentioned were possible and actual. I have painted these pictures for you because I have undertaken to furnish you sketches from life that you may recognize as correct, or at least approximately so, in your own county, in any county, in any state, in any land. You ought to know something of the world of men with whom you must soon deal, and "forewarned is forearmed." If I were to stop painting these pictures now, I would give you an entirely wrong conception of human nature. The good people, not wholly good, but people who are trying to live right lives and deal on honor, as in the sight of God, with their fellow men, outnumber vastly those who are parasites on society; not merely parasites, but foes to all that is good, and who are, by their crooked methods, sapping the very foundations of our civilization.

I shall not be able to find time to tell you of the scores and scores of nobler types of farm boys who are now making this nation great—the Brodheads, the Goodmans, the Wisemans, the Faithfuls, and dozens of other similar types that are well worth sketching. When I come to study the better class of farmers, and the businessmen of the great cities who have grown up on the farm, whose lives have been fashioned on farm models, the number of pictures that rise before me is so great that it embarrasses me to make the selection, and I find it impossible to find time or space to describe them all. They are not all perfect—none of them are. I hold it true that there is not "a just man upon earth"; that is, an absolutely just man, "that doeth good" always and everywhere, and "sinneth not," nor makes a mistake. I would do wrong if I were to describe such. The beauty and power of the scriptures rest largely upon the fact that they describe actual men and not perfect saints. I always take delight in reading that story of Moses when he, grown up in the court, the companion of princes, got mad and killed the Egyptian who was imposing on one of his poor and oppressed brethren. It was not right for Moses to do that, and he had to leave the country for forty years for doing it; but I do like to see a man's blood boil at the sight of wrong, even if he does make a mistake in his methods. If I were judge, these men would not get off easily. I have always felt more kindly to Abraham after reading that fib he told Pharaoh about his good-looking wife. It was mean in him to do it, and dangerous as well; but otherwise Abraham would have been such a perfect character that you and I would not think of trying to imitate him. The story of Jacob's sharp practice with Laban in dividing up the stock, where it was diamond cut diamond, the sin of David, and the foolishness of Solomon the Wise, all show that the Bible paints men as they are, and not as they should be; and in my feeble way I am trying to do the same thing for you. There is no man that I have ever seen that I would like to hold up to you as a perfect model. We are all but abridgements of a perfect humanity, and the very strength of character in one right line seems almost of necessity to involve corresponding weakness in some other line.

My dear boy, when I was your age I used to think that people most talked about favorably in the papers, and out of them, were those best worth knowing.[139] The preacher, the doctor, the judge, (my mother was always suspicious of lawyers) the members of Congress, were all great men in my estimation. I was disposed to look upon the very rich man with something of awe. You may, perhaps, have the same notions. As I grew older in years and experience, I changed my opinions somewhat. So will you. I found that the men best worth knowing, the men I could depend upon to stand by me in everything that seemed to them right and just, were not the smartest men, nor the richest, nor those the newspapers talked about, but the plain, common people, of whom there is usually very little public record beyond the fact that they were born, married, and in time died, leaving more or less of an estate, honestly gathered, to be divided among their wives and children. They do not, it is true, seem to make very much stir in the world, but if they were taken out of it, there would not be much left that is worth preserving, and the end of all things might about as well come at once. We could get along reasonably well with less than half the doctors, with one-fourth the lawyers, and we might even spare a few of the preachers. We could very well spare about nine out of ten of our small politicians, and might get along, in a pinch, without the millionaires; but we could not get along without the common people who rent from others, or own their forties, eighties, or quarter sections, or their modest homes in the cities; whose daily toil sweetens their bread, who live honest lives, train their families to habits of thrift and economy, and who form the sound and honest core of their church, their political party, and, in short, make this nation, and all nations, great.

The sooner you know these people and get in close touch with them, the better for you. They are the real source of what we call "common sense," which, outside of the Holy Writ, is the safest guide in all the affairs of life. If you are ever to retain permanently any position of trust and power that you may secure, and thus become a man of wide and commanding influence, you can do it only by being worthy of the abiding confidence of the plain, common people. You have heard, perhaps, of the advice President Lincoln gave to Governor Oglesby, as follows: "Stand by the common people, Richard; keep close to the common people." The deserved confidence which the common people had in Abraham Lincoln was the secret of his great power; and his ability to retain that confi-

dence in the most trying times this nation ever saw is the most convincing evidence of his supreme greatness of soul. It was the faith of the common people in Horace Greeley that made him the tribune of the people, and that gave the *Tribune* the regal power it wielded in those days, so full of peril to the republic. The common people will hear any man gladly who can at once teach the truth and live it.

Part Two
Letters to the Farm Folk

HENRY WALLACE
Reading Letter of Congratulations on His 78th Birthday

Uncle Henry's Preface to *Letters to the Farm Folk*

A good many years ago, I wrote a number of letters to the farm boy. I like farm boys, partly because I was a farm boy myself. I had then seen a good deal of life, and I wanted to help out these farm boys with my experience. I thought I could say something in a kindly, sympathetic way, that would help to start them right and keep them from making many of the mistakes into which they might fall through lack of experience. I have since learned that these letters have helped a good many farm boys. They have told me so themselves. Many of them now stand high on the farm and in colleges—are leaders of men, and they will have far wider leadership after a while as we older folk get out of their way.

Since I wrote these letters, I have had a good deal more experience of life, and it has occurred to me that I might broaden the field and write a series of letters to the farm folk—the girls as well as the boys, the fathers and the mothers, the old boys who have become men and have families of their own, and are bearing the burden of the heat of the day, the grandfathers and grandmothers, who have done most of their life work and who are now taking it easy and leading a less strenuous life, as becomes them.

The conviction has been growing upon me of late years that the biggest thing on the farm is not the land nor the livestock, but the farm folk, the people who live on the farm and out in the open country. These letters, therefore, will not be agricultural, but human. Do you know that the biggest thing in life, whether in city or country, is to be just a fine human being, interested in all things that interest or should interest human beings?

Advice for the Farm Parent

My Dear Folks:

Of all farm folks, the mother on the farm is the most important; in fact, she is indispensable. Who ever heard of a bachelor undertaking in blood earnest to farm for any considerable length of time, or a widower any longer than the conventions demand he should wait before taking unto himself a second helpmeet? If a man dies, his widow can carry on the farm better than a widower could. Sometimes, in spite of her lack of experience, she carries it on better than her husband ever did.

Upon the wife and mother fall the heaviest burdens of the farm, even though the husband does all a man can do to lighten them. She bears most of the double burden of rearing the children and making the home. To this day it is an unceasing wonder to me how my mother, who of nine children reared eight to manhood and womanhood on the old farm, managed to do it. It is also an unceasing wonder to me how my wife reared five children out of seven to full maturity in a little country town. Looking back, I can not understand it.

I wonder sometimes whether it was not easier, whether on the farm or in the town, to rear a family thirty years ago. To bear children, to go down into the valley of shadow without knowing whether she will ever return, to endure mortal agony, to care for the little one through all the ailments of childhood, through colic, teething, chickenpox, mumps, measles, perhaps scarlet fever or diphtheria; to all this with child after child through the productive period of woman's life; and all the while to keep the home with little or no help; to cook and bake and wash and scrub and iron, and keep the home as a sacred spot—surely the woman doing all this even reasonably well is entitled to a crown such as graced the brow of royalty, and to laurels such as were never bestowed on the victor at the Olympic games.

More than all this, to her belongs mainly the training of the children. They are hers in a more real sense than his. They are formed of her substance, bone of her bone, flesh of her flesh. They reflect and express her moral and mental moods during the nine months that preceded their birth. For the first few months, they live from her body. She eats and drinks for them. She produces for them the moral and spiritual atmosphere in which their natures expand. They reflect her smiles and her frowns. Father is simply a man who is there part of the time, whom they do not understand, and of whom they stand more or less in awe. His voice sounds coarse and harsh to them, even when with the best intentions he tries to soothe them to sleep with a lullaby. It is mother's kiss that makes them forget the pain of the first bump, or of their first acquaintance with a thorn or a fire. It is her voice that assures them they need not be afraid of the stranger, and that warns them of the danger in their efforts to get acquainted with this strange world into which they have been ushered without their knowledge or consent. It is mother who stills the strife between the little folks, who gently curbs the selfish desire inherent in the hearts of children; who teaches the first lessons of sacrifice and unselfishness—which we must learn in child-

hood if we are ever to learn them thoroughly, and if we do not then, must learn it through sad experience.

It is the mother who does all this, and is enabled to do it because of that undying affection for the children of her own body, that enables her to do for love's sake what she could not do or endure for any other consideration. To her, with her lesser strength and more delicate constitution, is given a power of endurance which we beings of greater physical strength and coarser fiber, mental and moral, must admire but can not either understand or imitate.

There are, however, two or three things that the father can do. He can so arrange the house that the steps necessary in doing the work will be the fewest possible. If it is necessary to reconstruct the kitchen and dining room, do so, and do it according to her plans; that is, if you really love your wife. You reorganize your barns to suit yourself, put in water systems to save labor in feeding livestock. Why not do as much and more for the mother of your children? Are the wives and daughters of less value than horses or brood sows or fat steers? We provide ourselves with the best labor-saving machinery on the farm; why not the best in the home?

Another thing you can do: You can manifest to your wife something of the affection which you lavished upon her in your courting days. I say "manifest." There is a great deal of affection among men towards their wives that remains unspoken. Men forget to continue to play the lover, and assume that having once convinced them of their undying affection, there is no need of saying nice things after the honeymoon is past. Women are not built that way. They hunger for words of affection and endearment and commendation, and hunger even more for the affection that flows from tenderness of heart, and all the more because it is an expression of the tenderness of the strong.

Another thing he can do: He can sustain the authority and the discipline of mother over children. The little rascals soon learn how to take advantage of her affection and love for them—and ask her to wait on them when they should be depending on themselves. The boys will try her out, not wickedly, but to see how far they dare go. It is then time for father to appear on the scene and say a few words or do a few things that will be worthy of remembrance.

Just here I'll make a confession: "Henry, if you don't behave, I will tell your father." These were words that I used to sometimes hear over sixty years ago. My father was a typical Scotch-Irishman, with eyes that seemed to me steel-gray when mother made her appeal for help in managing us, but melting into blue when we had pleased him. None of us cared to look into the steel-gray eyes, and none of us hesitated long when he said: "You do what your mother tells you!"

Uncle Henry

My Dear Folks:

Have you ever noticed the strong desire among middle-aged and old men—who were born and reared on the farm—to return from time to time for a visit to the old home? Often they desire to buy it, and spend the last of their days on the spot where they first saw the light. They seldom do buy it, for times have changed, and after a visit of a few days, they find that the old home does not appear and seem as they remembered it. The water in the spring does not taste as they thought it would. Strangers live in the old house, and the old furniture is gone. The neighbors wonder what brought them back, and what they see in it; and soon they themselves conclude that the particular thing they sought does not exist except in memory.

You seldom see or hear of anyone who is city-born burning with a like desire to visit his old city home. He can pass the house in which he was born, or in which his father and mother died, but with little trace of emotion. Why is this? Simply because the home life on the farm is different, and always must be different, from any other home life on earth.

The man in the city has his place of business entirely apart from his home. It may be a mile, ten miles, or twenty miles distant. The business itself has no connection with home life except in the way of dollars and cents. The wife and girls, and often even the boys, know nothing about the details of the father's business. They have an interest in its success or failure, that is all.

It is entirely different with farm life. The home and the place of business are one. The good wife is as deeply interested in the business as her husband. She watches the field of growing corn, wheat or oats with as deep concern as the man who sowed and cultivated it. A disastrous flood, hailstorm, cyclone, or drought affects her as deeply as it does her husband. Her mother heart responds to the motherhood among the livestock. She sympathizes with the cow mourning for her lost calf, with the brood mare that has lost her colt. She understands how they feel. The boys and girls as well soon begin to speak with an air of pride of ownership of the farm and all that belongs to it. It is "our" farm, "our" cattle, "our" colts, "our" home.

They all share in the work of the farm, each according to his ability. Their food, their clothing, their education, their pleasures, all hinge on prosperity of the farm. Their nights, as well as their days, are spent on the farm. Their association is mostly with other farm folks, kindred spirits. If trouble arises in the farm home, it is settled right there and then. If the boys have a misunderstanding with father, or with each other, mother is there with wise counsel. If the girls do not get along very well with mother, father can set things straight. But woe to the outsider who dares intermeddle in these family misunderstandings. There is loyalty to the family on the farm. There is loyalty to each other among home farm folks, such as is seldom seen elsewhere. The "gang" may rule or ruin the boys and girls in town, but on the farm the "gang" is the family itself—the natural gang.

Life among farm folks is not a life of ease and luxury. It is not quite natural or pleasant for the boy or young man to rise early in the morning, feed, water, and harness the team, eat breakfast by lamplight in order to be out in the cornfield as soon as he can see to husk. It is none the pleasanter if the husks are frosty, and the hands chapped from the husking of the day before. It is not the easiest or the pleasantest thing for the tired mother to arise and prepare this early breakfast, even if the fire has been built, the coffee ground, and the kettle filled and boiling. It is not the pleasantest thing in the world for the girl to be up before breakfast, wash the dishes, and be ready in time for school.

There are days that are very long and very hot, following the cultivator or shocking ripened grain. Farmers don't become rich with any alarming rapidity. There are no swollen fortunes made on the farm. It is a steady daily grind, with something for everyone to do, suitable to the age and the strength. Therefore, it is not much wonder that the boy or girl who has visited the town or city, or read of the luxury and ease of rich people, or of the sudden accumulation of wealth, becomes dissatisfied with what they call the drudgery of farm life, and breaks away, leaving the farm for what it seems at the time to be "fairer fields and pastures new." They know only one side of city life.

And yet these are the very men and women who long to visit and sometimes own the old farm, hoping to find there the real and lasting satisfaction they have sought in vain out in the wide world.

They have discovered at last that what they deemed sore trials and hardships in their childhood were the very things that developed the character that enabled them to achieve success. They know, too, that the city has its hardships and trials and tragedies fully as great as any in the country, and often much greater.

At the farmer's table there are no fancy dishes for the epicure, served on silver plates by trained servants keeping step to the music of the orchestra. Rather it's "Help yourself!" But there is a keenness of appetite and relish, the result of exercise in the open air, for which the epicure vainly longs. Water from the wellspring, whether in the gourd dipper, the glass at the table, or the jugs in the shade of the field, has a finer tang than Appollinaris or Vichy.

Remember that Midas must in time live plainly, more so than the farmer, if he is to live at all. I said to a very rich man the other night: "Tell me what you eat." He replied, "I eat an egg and some toast for breakfast. I eat no lunch at noon. I eat no meat except on Sunday, when my children visit me, and then have chicken or turkey. I eat a very light dinner in the evening." Then I said: "In order to live at all, you live plainer than the farmer or the laboring man. It's either that or nothing at all?" "That," he said, "is very close to the truth."

Nor is the home of the farm folk a stranger to sorrow and trouble. An accident may occur. There is broken limb, which calls for the sympathy and active help of all the family and the neighbors. Mother is taken sick, and you are sent posthaste for the doctor. You find him leisurely eating his dinner. You urge haste, and feel that he is unnecessarily slow. He is used to such calls. At last you get him out to the home. As he comes out of the sickroom, you read serious trouble in his face. The sunshine then loses its brightness, and nothing looks bright to you. The very face of nature is changed until there is a change for the better in that sick room. You never knew before how closely her life is bound up in yours.

The little brother or sister sickens and dies, and for the first time you look upon the peaceful face of the dead. You will never forget that. Worse still, a sister fixes affections on a scoundrel, and you know it; but what can you do? If you could thrash him within an inch of his life, you would feel a lot better, but there seems no way of turning her affection from him. Yet you love her nonetheless. These troubles and sorrows bind you more closely in spirit to those that remain in the home, and make you more tender, more courageous, more truly human. It is part of the discipline of life among farm folk.

This life on the farm may seem to you a very plain and prosaic life, not at all like city life, nor even like the country life you see described in novels or magazine articles. All real life is more or less prosaic. All life worth living has its labor, its trials, its joys, its sorrows, and at times its tragedies. The life of the farm excels all other life in this, that it is really a family life, combining the home and business, and in which self-interest binds the family into a unit. It is a life in which there is useful work for every member of the family, from the youngster who can help in any way, to grandfather, who loves to do chores, and grandmother, who sits by the fire, knitting. It is free from the stupid conventions of city life. No mask is worn in the farm home. Farm folks are what they seem to be. When neighbors drop in, or they visit neighbors, they are just what they are. It is this freedom from conventionality, this openhearted, free-handed hospitality, that appeals to the city man, whether born or reared on a farm or not, and makes him long for a quiet, peaceful farm life.

Uncle Henry

My Dear Folks:

Whether farm boys or girls land in afterlife in the scrap heap, whether the hobbledehoy and his sister pass through the most critical period of life without damage or even blemish, depends mainly on the home itself, but very largely on the social life around them; and this depends very largely on the permanence of the farm home. Someone has said:

I never knew an oft removed tree,
Nor yet an oft removed family,
That throve so well
As one that settled be.

This is quite as true of the moral, social, and spiritual side as of the material or financial.

In my last letter, I talked to you about the home life of the farm. I am now speaking more particularly of the social life of the farm folk. We are social beings, made so, and we can't help it. If we try to help it, we become dwarfed and twisted both mentally, socially, and spiritually; we become narrow, selfish, suspicious, and comparatively useless. Man was made for the society of his fellows, and

without them he can never be fully or finely developed. Our children are born so, and they hunger for the society of other children, more than we older folks do for that of our fellows. Parents have each other for company, but that does not fully satisfy them except during courtship and a little while after that. Then they begin to hunger for the fellowship of others, the touch of a neighbor's hand, the sound of a neighbor's voice, or even that of a stranger.

Children are strangers in a strange world, and must learn it—not merely its geography and its animal life, but its human life. Notice how sharp are their eyes and how wide open their ears, when a stranger comes into the home and tells of things of which they have never heard before. "Little pitchers have big ears." Notice how greedily they devour tales of adventures in strange lands, how fond they are of stories, whether oral or written, what vivid imaginations they have. Ah! If we could but retain the keen perception of childhood, all knowledge would be our province, and we would be masters of our destiny.

The character of the neighborhood in which you live, the school your children attend, the plays and sports in which your children engage, become of vastly more importance than the kind of clothes they wear or the food they eat. But how can there be a desirable and satisfactory social life if there is constant change going on among the people of the neighborhood?

Farmers are in some respects a rather curious folk, differing from the man who makes his living through studying men instead of crops. The businessman's long experience enables him to classify the men he deals with largely into groups, and to deal with them according to his best judgment. He touches most men only on their business side, and need have nothing to do with them socially. It takes time to establish lasting friendships among farm folks. We take the stranger who has come into the neighborhood, whether as owner or tenant, by the hand—but we don't at once take him to our heart. We want to know something of his past, of his forebears; in other words, of his pedigree, before we encourage our children to be intimate with his, or even be intimate with ourselves. The experience of life, as well as the reading of history, has taught the wisdom of that advice of Polonius to Laertes in *Hamlet*:

> Those friends thou hast, and their adoption tried,
> Grapple them to thy soul with hoops of steel,
> But do not dull thy palm of entertainment
> With each new, unfledged comrade.

Today about forty percent of the lands of the corn belt are cultivated by tenants; and these seldom stay long enough to be assimilated into the social life of the community. Every one of my readers knows the extents of these changes; and every change tends to disintegrate the social life. These tenancies are the result of the moving of farmers to town, of the movement of the old, settled families to distant sections, of the purchase by capitalists of land as an investment, and for speculation by men who believe that the price of lands in the corn belt will go on increasing as they have for the last fifteen years.

The country suffers least if, when the farmer who has been a pillar in the country life retires, or, in the other words, moves to town, his place is taken by a son or son-in-law or nephew, who is a member of the community. We are acquainted with him, and are glad that the life of the whole family is not lost to the community. But on losing this older man, his worthy wife and unmarried children, we are conscious that something has gone out of our life that we could not well spare. We know their ways, their prejudices, and had unconsciously adapted ourselves to them, looking good humoredly at their eccentricities. We know their sterling worth. We remember their kindly helpfulness in times of trouble, their words of sympathy when loss or sorrow came to any of us, the help they have been in various ways to the entire community. They leave the country poorer for their going.

If a strange tenant comes in, we watch him and his. We don't know them, and they don't know us. We watch to see how he will farm. We mourn when we see him plow up clover and bluegrass pastures and convert them into cornfields, and there is no lowing of cattle in the empty barns or stables, but robbery year after year of the soil fertility which our old friend has stored. If he has a five-year lease at a moderate or high rent, we know the look there will be on our old friend's face when he returns to visit the fields he has left.

Or the farm may have been sold to an investor in city or town, faraway or near; and the incoming tenant is a stranger. We know nothing about him or his forebears, or his past history. If we find he has a one-year lease, we scarcely care to know him. Even the pastor does not know him; the Sabbath school teacher does not know him or his children. He is perhaps shy and reserved; his wife even more so. He doesn't know us or our ways. He feels shy about going to church, especially so if he happens to have little of this world's goods. He may have been a hired hand, and has scrapped enough together to make a start as a tenant; or he may have once owned a farm and lost it, and is forced to become a tenant. If adversity comes to him, such as sickness or death in the family, we then turn to him, and may be surprised to discover that in our neighbor we have found a jewel, or it may be something else. But even if our old friends have left us and gone into Canada, or California, or Texas, and the new occupant is in every way his equal or his superior, a better farmer, with the capacity for being a better neighbor in every way, we don't as yet know him. It will take time for us to learn to know him, time for him to grow into the country life.

These changes are going on, as every one of you folks know, in almost every neighborhood. They have been going on very rapidly for the last fifteen years, ever since the phenomenal rise in the price of lands began; and will go on as long as land in the corn belt section continues to advance. When that point is reached where sales become less frequent, and trading in or selling equities begins, or forced sales follow, there will be a readjustment that will result in still further changes, especially in the newer regions.

Meanwhile our county churches are being so weakened that many of them have been and are being abandoned. Our schools will either diminish in numbers or change pupils. Cooperation, in which lies the main hope of the country community, will be delayed. Except in favored communities, our children will lack the intellectual training that comes from a fine social life. What can you do about it? Not much just now; but something must be done, and that in thinking rather than doing. In my next letter, I will try to open up a line of thought on this subject which I hope will receive your serious attention.

Uncle Henry

My Dear Folks:

In my last letter I tried to point out clearly that the lack of the social life we so earnestly desire for our sake, and for that of our children, is largely due to the constant changes in the farm neighborhood. Some of you may be inclined to think that because this constant changing has continued as far back as you can remember, it therefore belongs to the natural or normal condition of country life; that we must put up with it and the best we can in spite of it. But it is not so in older countries, nor in some sections of our own; nor will it always be so in the corn belt. Ireland has a stable country life, because, while the young folks leave, the families remain. England and Scotland have stable country life, although more than four-fifths of the land is farmed by tenants. The same is true in France, where the land is mostly owned in small tracts by the farmers, and there is but little tenancy. Therefore, a stable country life is possible either under tenancy or ownership, provided the tenure of the land is practically permanent. The constant change in community causes the trouble.

When we get down (or rather up) to a stable country life, we will have a far more desirable country life than the foreigner, for we are not cursed with the social distinctions which prevail in these older countries, which, except in very rare cases, prevent a man from rising into a higher class than that in which he was born. We have, it is true, absentee landlords, as they have, but the tenant here does not have to look up to his landlord as an altogether superior being. "These are disgusting times," said an Irish landlord to me once. "Time was when the tenant did not dare to come into the presence of the landlord without putting his hat under his arm, nor sit down until he was invited to do so. Now the beggars come in with their hats on their heads, and sit down as if they owned the place. They have no respect for their betters."

How are we to prevent these constant changes which go back to the root of our trouble? Some of them will cease of their own accord. The time will come, if, indeed, it be not here already, when capitalists will not speculate in land. The time will come when a thousand dollars will buy as much actual value in farmland in the corn belt as anywhere else in the country, and the farmer will be satisfied to remain. Many intelligent men think that time is almost here even now. The time will come when farmers who find the conduct of the farm too heavy for their years, will find it suits them better to build another house on the farm, so that they can spend the rest of their days among their old neighbors, and not break with old ties by moving to town. Some have found this out already. The time will come when investors in farm property will find that it does not pay, by exacting a high rent, to sell their farms piecemeal.

The best way, however, is by law or custom or public opinion, to compel the landlord to do justice to the tenant, and the tenant to do justice to the landlord, and both to do justice to the land. The English Parliament has done this. This will make it to the interest of the landlord (and we will have landlords for at least another generation) to make the tenant as much of a fixture in the neighborhood

as the landowner himself. Death in time will take both, but we can submit with good grace to the orderings of providence.

At present the law allows the tenant to rob the land, or, in other words, to starve it. The law would put the tenant in jail if he starved his horses or his cattle, but we allow him to starve the land. The law would put the landlord in jail if he confiscated the horses of the tenant, but we allow him to confiscate the fertility which the first-class tenant stores in the soil, and seem to think it is all right. The law would put the tenant in jail if he sold the personal property of the landlord, but we are likely to approve of his robbery of the fertility which the retired farmer had stored in the soil when the farm was his home.

The state must begin by amending our landlord's lien law, and take away from the tenant his power to pledge everything he has in the world, except his wife and children and furniture, to the landlord for rent, or to pledge himself not to sell the crops (with which he is to pay the rent) before he pays it. This, however, is only a beginning, but possibly all that can be done at present; but we must not stop there.

What next? Can legislation do anything? Well, you and I are not legislators; but if we were, and had the wisdom of Solomon, we could not do a thing unless we had public opinion behind us. Even though we are not legislators, I am doing something, and you can do something, to mold public opinion, at least in your own community. You can have a very definite opinion as to what is justice to the land, the landlord, and the tenant. You can think this over by yourselves. You can talk about it to your neighbors, with those who own land and those who rent it. Some of us are landlords; some are tenants. We can devise for ourselves a method of renting land which will make it to the interest of the good tenant to stay, and of the landowner to have him stay indefinitely. When we have done that, we have solved the problem and put a lasting foundation under country life, and shall then have a social life in the country that is worthwhile.

The English government has solved the problem in Scotland and England by compelling the tenant to put back into the land the manurial equivalent of the grains he sells off it, by preventing him from selling straw and roots, which must be fed to livestock on the farm; by compelling the landlord to pay the tenant for the manurial value of the foodstuffs he has purchased and fed to livestock, or else let him stay until he has used up this fertility, and also by forbidding the landlord to raise the rent because of improvements the tenant has made. Hence, in these countries it is not to the interest of the landlord to change tenants, nor of the tenant to leave. In the meantime, the land, which feeds the nation, is itself fed. You and I know how easy it is to do the decent thing, when it is to our interest to do so. When duty and interest pull together, what a magnificent team they make! When they do not, what uphill work it is!

But, you say, this would make us all stockmen. Well, that's what we ought to be, and will have to be sooner or later, if we are to have any satisfactory social life in the country. Growing grain for sale off the land starves the soil. I am speaking now for the voiceless land. It will not feed you unless it is fed; we will then become poorer and more discouraged; and how can we have any satisfactory social life among poorly fed and discouraged people?

Uncle Henry

My Dear Folks:

As I look back over my boyhood days, sixty or more years ago, and I think of the change and temptations incident to boy life, and the fatal mistakes I might have made, I get solicitous about the boys growing up on the farms and wonder whether I may be able to say a word to both boys and parents that may be helpful in what I regard as the greatest of farm problems, the rearing of splendid human beings. For, in the last analysis, this is the real work of the farm, to which everything else—the growing of grains and grasses, the rearing and feeding of livestock, the buying and selling—should all contribute.

When that boy of yours was born, when you looked into his wondering eyes for the first time, you were a very proud and happy man. Is it not so? Can you ever forget that day? I can't. To your friends you seemed an inch taller, or at least half an inch. You had a smile on your face that no one could mistake; your shoulders were straighter, your hand-clasp firmer and more cordial. It would be a very ill-natured man who could pick a quarrel with you that day, or even the day after. The proud mother folded the boy to her breast, the fountain of life for it, and felt that through the very gates of death she had entered into a larger life, and had full and abundant recompense for her suffering. You felt just a little different toward her, and she towards you, than ever before. There was a new and

intimate bond between you now. You really felt, and so did she, whether either of you said so or not, that he was altogether the finest and most remarkable baby that ever was born. I don't blame you, for I have been there.

Well, the boy has grown up, is particular about his neckties and the cut of his clothes, and you have been studying him all these years. You have been wondering what you were going to make of him, or, rather, what he was going to make of himself. Whatever that may be, you are anxious to get him started right.

A right start means more than you realize. It means a sound body; it means an active mind; it means good habits, and it means preparation for the life work for which he is best fitted. If you have fed him with plain, wholesome food, if you have set him a good example and high ideals, if you have trained him to habits of industry and economy, the rest is comparatively easy. If not, you and his mother have a hard task before you; for these are the very foundations of character.

They can not all be farmers or farmers' wives. There are too many of them for your acres, but you want at least one to be a farmer, and the problem is how are you to start him, and all the rest of them, right. From one point of view, the most important part of the boy is the head; for if we get the head started right, the rest will follow. The most important lesson for the head—and this should be indicated from the very beginning—is that you are one family, the most important cooperative organization that was ever formed. The slogan for the family should be: Each for all, and all for each. In the battle of life, the members of the family must all stand together, and, if need be, help each other against all the world. This will greatly simplify matters when you come to find out what each is adapted for; for you will join in giving him the right start.

You are wondering whether John will make a farmer. He is helping you in the farm work as best he can; so do all the rest—of course they do. Is his help mechanical, a mere matter of duty? Does John really like farming, or does he regard the work as drudgery? How are you to find out? Get some good farm papers—agricultural, livestock, dairy, horticultural. Don't urge the boys to read them, but keep them lying around handy. Notice which one reads them for himself, and what part he is most interested in.

If John takes an interest in corn growing, tell him you will give him an acre. Let him have all he can grow on it for his very own, and tell him that if he grows over fifty bushels on it, you will pay him twice the market for the excess. If either John or Tom or Jim is interested in articles which deal with breeding, rearing and feeding of livestock, give him a pig or a calf or a colt. When it is sold, be fair with him and give him the money. If you need money badly at the time, ask him to loan it to you at interest, and give him your note.

Notice all along what kind of books the boys read, what kind of papers. Don't tempt them with cheap, trashy magazines, or with papers that come in uninvited, and keep coming after the time is out. Protect your boys from worthless reading. Read occasionally a copy of every paper that comes into the house, and if you find anything that might taint the mind of boy or girl, burn that copy at once and order the publishers to discontinue it.

The farm boy is pretty apt to regard farm work as drudgery. Well, don't be a drudge yourself. The way to transform drudgery into pleasure is to get the mind interested in the work you are doing, not the attention merely, but the mind itself. Get interested in the laws of nature, which rule even in such things as digging ditches, hauling out manure, milking cows; for example, how water gets into the ditches from the sides or from below, and why; the wastes of manure heaps; why manure is more efficient in some kinds of soils than others; why cows give more milk on some kinds of feed than on others. No matter whether the boy is to be a farmer or a preacher or a doctor or a businessman, this kind of training will be a mighty help to him in afterlife.

No mental equipment, however, will start the boy right unless he has right ideals, and these are the result largely of the home and home training. If you have not taught your boy from childhood up to reverence his mother and be the guardian and protector of his sisters; if you have not taught him that it is his duty to give everyone a square deal; if you have not taught him to be honest, not because it is the best policy, but because it is everlastingly right, the mental equipment you have given him in the school or college, or that great school, the farm, will not start him right. For, important as the boy's head is in fitting him for life, the heart, his ideals, his attitude towards God and his fellow man are far more important.

The well-ordered farm home is the best and safest of all places to start the boy right. Every bit of right training he gets there will be helpful to him in afterlife, no matter what occupation or business or profession he may follow. Whether the boy at whose birth you were so proud and happy will be an honor or a dishonor to you depends

largely on the start you give him in the home life. It is not the
financial start, but the mental and the moral and the spiritual which
determines success, and this is as true of the girl as of the boy.

Uncle Henry

My Dear Folks:

Farm folk are just like any other folks. They have bodies to be fed
and clothed, and minds that have to be fed, developed, and fitted for
the work of life. What kind of clothing we should give our folks for
their bodies depends on age, sex, the kind of work, and the income
and means at command. What kind of food we should give them for
their bodies depends on the same conditions.

The same is largely true as to the kind of brain food. We feed
the baby with its mother's milk. By and by it gets its first tooth, and
there is great rejoicing in the family over that event. Others follow,
showing that the child is getting fitted for digesting the food of the
family after a while. So children need a different kind of brain food
from that of older folks. In both physical and mental nurture, the
food in infancy must be given by the mother.

The child loves stories, stories such as no one but mother can
tell. It is getting its mental food largely through the senses, and it
gets its education in the same way. The child at five years old has
learned more per year than it will ever again learn in its life, even
if it lives to be as old as Methuselah. In childhood, the imagination
is vivid, the powers of observation keen. The child loves stories,
and never gets enough of them, as some of us know from their
own experience. These are food, palatable food, to the child. Solid
teaching, such as delights us when older, is medicine to the child.
Children never tire of the story of Moses, of Samuel, of the infant
Jesus, nor do they ever tire of folklore or fairy tales, or animal
stories.

We have therefore put, without hesitation, into the hands of
children the story of Jack the Giant-Killer, Jack and the Beanstalk,
Grimm's Fairy Tales, and Hans Christian Anderson's Fairy Tales.
These tales belong to the world. They have entranced children in
India, in China, in the tents of the Arabs, as well as in the palaces of
princes born to the purple, and in the humble homes of the peasants.
Don't make them wise before their time about Santa Claus. Let

them indulge their imagination in this. They will learn the real truth by and by. It may be humiliating to us older folks, but children are sharper than we are, and get to the kernel of things.

The time of childhood is past, and we must send our children to school. We don't send them to school to acquire knowledge, except in an incidental way. We send them there to learn to read, to learn to write, to learn (as we used to say) to "figure," to learn to reason and to draw conclusions. In other words, we send them to school to acquire the tools that will enable them to read intelligently what others have written, that will enable them to put this and that together and draw a conclusion. (On this subject I shall have something to say after a while.) When a child has learned to read and write and speak its mother tongue, the whole world of knowledge will be open to it. If it has learned to think and reason correctly, it may take, as Lord Bacon said, "all knowledge for its province."

As adolescence comes on, children will read love stories. It is not a question of whether they should read love stories or not. They will read them whether or no, and boys will also read tales of adventure. The question with us is: What kind shall they read? Shall they read something which gives them accurate and true views of life, or the opposite? There are certain kinds of love stories which should be absolutely tabooed, because demoralizing in their effects, setting forth utterly wrong ideals. In this class we might safely include all the stories in the cheaper magazines, and some of those in respectable higher priced ones, which, in the best language and with the finest illustrations, on good paper, convey the impression that young people can go to the limit of actual sin and yet escape. They don't escape. Their minds are poisoned. Impressions are made from which they will recover with the utmost difficulty and self-loathing.

We would also exclude a large class of so-called Sunday school literature, in which the good die young, in which negative characters with negative virtues are pictured as being normal and desirable, in which isolation from the world is often presented as the highest type of virtue. Farm children have red blood in their veins, and their reading should teach them that health, vigor, energy, are really desirable things, that the highest type of courage is in defending the weak and helpless; that the hero is not only the man who slays the enemy in battle, but that conquers the evil in himself and stands forth as a champion of right. Man is naturally a fighter. We should teach our boys to fight, not against each other, but against evil within themselves, and the outrages which boys in wanton cruelty commit on each other. That's the type of courage we should

cultivate, that and defending the weak against the strong.

In the reading you put before your boys, you should teach them what kind of a world they live in—books telling them about different countries, history, especially the history of their own country, the history of great men—not that of Napoleon, but that of Lincoln. Boys are naturally hero worshippers, and parents should shape that hero worship by giving them books to read which will develop the highest ideal and stimulate them to live up to it.

We have just now a great wealth of reading matter both for boys and girls which is food and a healthy tonic to the young. There is also a lot of literature, so-called, which is deadly poison. In nothing should farm folk be more careful than in this matter of keeping out of their homes bad literature, tainted reading matter, anything that is doubtful. Parents should read the literature their children will read. They should read the magazines which come into their homes, and determine for themselves whether they are likely to develop the type of character they want to see in their young folks. Better still, let the children read aloud to you at least samples of what they love to read.

You should read a great deal about your own state and its government, and about your own local government, so that you may stand for the right men and right measures. You should read daily papers of only the highest type. If others come your way, read by headlines only. One of the best ways to determine the quality of a paper is to look over the advertisements. If you find patent medicines galore, get-rich-quick schemes, things to be given away, whisky and cigarette advertisements, and all that, keep that paper out of your home. In judging papers that are fit to come into the home, read the editorials. If they are thoughtful, earnest, that paper is probably not edited from the business office. If they are weak, without courage to fight against evils, that paper is a mere moneymaking concern, and should not be encouraged. Some have a few excellent articles, apparently to sanctify the other pages, even as you will find some religious papers with advertisements that a decent agricultural paper would not insert. Stop both! Every one of you should read about the business of the farm. The time has gone by when farming is a matter of "brute strength and awkwardness," a matter of luck. It is science and art combined. You get the science from books and papers and bulletins and lectures. You get the art by practice. If the science, so-called, won't work out in practice, then it is not true science. Our attitude in reading should not be that of a receptacle into which things are poured, not that of an animal that is being drenched, nor that of a calf that is fed by hand. "Only that good profiteth, which is taken with relish." That is, only that good profits you which you enjoy and which you can assimilate.

In reading anything, we should read with an inquiring mind, and then ask ourselves if it is true, if it conforms with our experience and observation. The agricultural colleges and experiment stations, and the department of agriculture, are furnishing us with a vast amount of valuable information. There is no dearth of it. We must remember, however, that these people are not the fountains of wisdom. They often make mistakes, sometimes serious mistakes. They need to revise their conclusions. What is scientific truth may not be definitely and firmly settled. Scientists are often merely feeling their way toward the real truth.

Every farm should have a library, bought one book at a time, and read carefully. In buying books for myself and my home, I avoid the bestsellers of the day. They are hardly worth the shelf room. After they have been out two or three years, and are still selling well, it may be worthwhile to buy some of them. The bestsellers are aimed at the general public, and strike a lower level, and seldom give one anything worth thinking about. Of course there are some exceptions.

Of agricultural papers, take only the best, and pay for them yourself. Don't read a paper that throws in a lot of junk to get you to subscribe to it. Such papers are not worth your time to read them, and the fact that something else is offered to sweeten them up, shows the publishers themselves don't think them worth reading. Every farmer who makes any profession of Christianity should take a religious paper—two or three of them, for that matter—one of his own denomination, one of the aim of which is to reach good people regardless of denomination. He should read them without bias, and endeavor to ascertain what is the truth.

The best book for any and all of you to read is the Bible. No other book like that has ever been published. It has history, poetry, philosophy, ethics, inspiration, something that puts you on your mettle, something that quiets you, something for comfort. It tells among other things of the faith of the farmer in sowing his seed. It is written mainly by farm people. Many illustrations used by the great master of all of us are drawn from farm life, whether he was a carpenter or not.

You see, I have not told you what books and papers to read, and I don't intend to, because reading must be different for different people, and you must be your own judge of what to read. Be careful to judge wisely.

Uncle Henry

My Dear Folks:

I have talked about the friction that often arises in farm homes from one cause or another. There are, however, often more serious troubles than this. In the course of the year we receive numbers of letters, mostly from boys and girls who are alienated from their parents, who think their fathers and mothers have not treated them right, and they ask us what they should do. For example, a boy of seventeen wrote us that he did not think his mother gave him enough pocket money for the movies, suppers, and dances. Another boy of about eighteen wanted to know what we thought his father ought to pay him for work on the farm until he was twenty-one years of age.

Neither of these boys seemed to realize that their parents did not legally owe them anything except food, clothing, and education; that their time until they are of age legally belongs to their parents, though their parents are obliged to support them, and are liable for their debts.

Now and then, when a boy begins to assert his rights to liberal spending money, or a girl to extravagances in dress, or having what some girls and boys call "real pleasures" in life, it is necessary to inform them that the law gives them no rights to even what they earn before they are of age, while the law of the state puts the parents under obligations to feed, house, clothe, and educate them according to their means, nothing more. When this idea is firmly fixed in the mind: that the favors they get are not matters of right, but expressions of love and affection, it is then the proper time to distribute favors, such as the parents can afford to give. Spending money does not come as a right, but as a gift, an expression of affection; and it should be given as freely as circumstances allow so long as the young folks are disposed to use it wisely. In fact, every boy should be trained in a wise use of money before he comes of age, and every girl should be trained to buy, and, as far as possible, make her own clothes. It is part of her education.

The father, however, who stands on his legal rights with his boy or girl, except when it is absolutely necessary to clear the vision of that boy or girl, is almost certain to find that he has made a mistake. He has, it is true, his legal relation to that son or daughter, but he has also the relation of parent, which altogether overshadows the legal relation. He should always be ready to do for the boy far more than he is legally required to do. The boy, because he is his son, naturally feels this, and the failure to do what nature, or rather, this relationship, involves often leads to differences in families which may be lifelong.

I have in mind now a farmer who owns a good farm worth thirty thousand dollars. He has a son who is anxious to secure a better education than the country school can afford. The father had only a common school education. He has got on in the world by the advance in the price of land, due, not to any efforts of his but to the rapid growth of a nearby city. He has no sympathy with the boy's aspirations. He grudgingly gives the boy his time during the winter months when he does not need him at home. The boy comes to town school, works for his board, tends furnaces, etc. He is an earnest student, but his father insists on his return to the farm as soon as spring work opens. Will that boy have the feeling for his father that fatherhood should involve. I would really like to know what the boy's mother thinks about it. Farm women rarely write me about their troubles. Most of them bear bravely and silently the real sorrows of life. I know that in all these cases there is another side, a side which I can not know. Hence I can not advise, much less judge.

Many of the real troubles that arise in families, however, would be avoided if each parent would learn to control his temper. There are every day many things about the farm that try men and unless they have learned to control their tongues as well as their tempers, there is likely to be trouble in the home. If a boy is sent out with the cattle, and is told to put up the bars, but, becoming interested in chasing a squirrel or rabbit, forgets to put up the bars, and the cattle get out, break into a neighbor's cornfield, and two or three of them are foundered, it requires a good deal of grace for a man to hold his tongue, or at least to keep it off the careless boy. Yet the boy really did not mean any harm. He was just careless, as most boys are, and forgot.

You must not take the hasty words uttered by either parent or child, especially if uttered under great provocation, or when tired or worried, as an expression of their real character. We are any of us liable to fly off under provocation, and nowhere is there more

provocation than on the farm. What is said expresses only the pass-ing mood of the moment. Many a father who will scold his boy and his girl shamelessly, and, we might say, brutally, has in his heart a deep and abiding love for them, and would do anything in the world to promote their welfare. This may not excuse him, but we would do well to remember it.

We find, here and there, a man who takes pride in giving a boy or a neighbor a "piece of his mind." He seems to think it is an evidence of manliness. It is nothing of the kind. It simply means the lack of self-control. This lack of self-control is one of the causes of trouble, and sometimes alienation, in farm families.

This alienation and trouble is no new thing. Sometimes the fault is with the parents. If Rebecca had not made a pet of tricky Jacob, who no doubt looked after the stock fairly well, and was a sharp trader; and if Isaac had not thought so much of the good-humored, careless, even reckless Esau, who was fond of hunting and knew how to cook a dish of venison to Isaac's taste, the history of that family would have been entirely different. If Jacob had not made a pet of Joseph because he was the son of the only wife he really loved; if he had given credit to the boys who really did the work on the farm and looked after the stock; or if he had been wise enough to conceal his favoritism, which he really could not help, the whole history of the family would have been different. It is true that the Lord overruled this folly, as He does ours, but it was folly nonethe-less.

I have called attention to the impossibility of treating all chil-dren in the home alike, because they are not all alike. We are not alike to our friends. Each friend draws out in us certain traits, and we can't help it. This fact complicates very greatly the problem of raising a family. But fathers sometimes exercise their authority in a very unfatherly way, and mothers sometimes show favoritism to the boys as some fathers do to the girls, in such a way as to lay the foundation for the future alienation. I have known men and women who never got over the feeling that their parents did not treat them fairly; and very frequently the parents are to blame. More frequent-ly, however, the children are to blame.

Boys are very apt to be conceited; girls are very apt to become vain. There is a certain age when a boy thinks he knows more than his father. I confess that I passed through that stage myself, and I have had to laugh many a time since at my own conceit. I have never forgotten the peculiar smile that passed over the faces of fa-ther and mother, when I was suffering this egomania. Is that a new word? Well, it simply means an exaggerated idea of one's own importance and wisdom. I now interpret that smile as meaning: *Well, that's a rather amusing, son, but you'll get over it after a while. You will know better, and be ashamed of this nonsense*—all of which was a true prophecy.

If fathers were wiser in exercising the authority which fatherhood gives them, and would gradually take their boys into their confidence, talk to them as younger brothers, give them the counsel and advice which their age and experience should fit them for giving, we should have less alienation in farm homes, and in city homes as well. For it is even more difficult to grow a boy or a girl under the temptations of the city than in the quietude of the farm, and to keep them devoted to the home.

I think we all make a mistake in not carrying our outside manners into the home. We sometimes forget that the family is just as deserving of polite treatment as the stranger or the neighbor. There has been many an outbreak of ill-humor on the farm that has been quieted all at once on the approach of some stranger or a neighbor or friend for whom we feel deep respect. If we can control our tongues in the presence of a stranger or neighbor or friend, why can't we control them in their absence? If we all lived under the feeling that "Thou God seest me," there would be a measure of politeness in the homes of both country and town such as we have never seen. If we wrong a neighbor, we beg his pardon. If we "speak unadvisedly with our lips," to our children, and do them wrong, it should not be beneath the dignity of any father to say to his boy: "My boy, I was wrong in that. I beg your pardon." It will greatly develop the manhood of the child when he has done wrong, if he has been taught to go to father and mother and say: "I've done wrong. Won't you please forgive me?" There is nothing that will so guide and elevate in character as the cultivation of feeling that there is one above who hears every word, knows every emotion, sees every action. We would have no trouble in the government of any city, if there was a policeman in the shape of an active conscience inside every citizen. There would be no trouble in society and none in farm homes, if we all realized that we are ever under the all-seeing eye.

It is well for parents to remember that the children did not come to the home of their own accord; that they had nothing to do with the selection of either their home or their parents. Coming thus without their own consent, their parents are under obligation not merely to give them proper food and clothing, but to make the home as nearly a child heaven as they can. The mother is quite likely to do this according to her best light; for she has a degree of natural affection that the father as a rule has not. It is the affection of the ewe for the lamb, of the mare for her colt, plus the intelligence of the human being. The father who does not shield and protect they mother, and does not uphold the authority that naturally belongs to her as a mother, lacks something of being a real husband and a real man.

Do not forget that the American people will be largely what their home life is. If the child be cowed and abused until his nature is dwarfed and twisted, until his spirit is broken, he is apt to deal with others in the future as he was dealt with. The American homes are the cradles of our liberty. If the home life is right, if the training is such as to develop loyalty to God and to the home and the family, there will be less trouble in state and nation. If we see ugly traits cropping out in our children, it is well to remember that they are our children, and that possibly some of those traits are inherited from us, and that we are in a measure responsible for them, and it rests with us to help our children overcome them.

Uncle Henry

Advice for the Farm Couple

My Dear Folks:

As I am presuming to talk to you sympathetically, and in a measure of confidentially, about every phase of your life, I can not well avoid talking about possible friction in the farm home. And when I say "friction," I mean friction, and nothing more. I do not mean dislike or enmity, much less quarreling, but simply friction, the lack of perfect adjustment to make everything run smoothly.

I was coming home on a limited train after a month's absence. Everything had gone smoothly the whole month through. Everyone I met appeared glad to see me; my friends did everything they could think of to make it pleasant for me. Every train was on time till the last day. In the afternoon, we changed engines for the final run home. When we got behind the new engine, I soon saw that we were losing time, stopping where we did not expect to stop, and stopping longer than we expected. Finally I asked one of the train-men what was the matter, and he answered laconically: "Hot box." There had perhaps been a little dirt on the axle where the wheels turn; hence friction, heat, then it took fire, and we had to wait until it cooled, and more grease was put on. We limped along, changing engines twice, getting home four hours late, and all the passengers tired and irritated.

You may have forgotten to grease your buggy, or perhaps to clean off the end of the axle and grease it properly. You hitch up to go to town in a hurry. Before you get very far, one wheel does not seem to work right, and by and by it won't work it all. You get out and examine it, and say in disgust what the trainman said: "Hot box!" A little bit of sand; that was all. A little bit of friction often creates a lot of trouble.

No two persons can get together to form that most blessed thing on earth, a happy farm home, without there being need of more or less adjustment, even if they are deeply, and as their friends sometime say, foolishly in love with each other. There will be some point of friction, needing adjustment and forbearance, and it will be apparent by the time the honeymoon is over—sometimes before. Sometimes it appears even before marriage, in the shape of a lover's quarrel. Some little difference in taste or opinion, some habit which has no moral quality whatever, will create feeling, by and by heat, and then flame.

Generally speaking, the younger that people marry after reaching full growth and sufficient maturity of character and knowledge of the world, the easier it is to make the necessary adjustments. The longer marriage is delayed after that, the more fixed habits become, and the more difficult of adjustment. When the marriage is the old-fashioned love marriage, adjustment usually will be made, and with secret pleasure to the adjuster, but if it is a marriage made solely for money or position or a home, there is likely to be friction and serious trouble. It is well to be sure that the proposed marriage is really a love marriage, and not a sudden outburst of sexual desire, or the result of a disappointment on either side. Natural affinity, or magnetic attraction, or whatever you may call the attraction of maid for man, has its place, but if there is not back of it all a mental and moral affinity, which lays the foundation for a lifelong friendship, there will be friction, good and plenty. Much of the happiness of married life lies in the skill of both sides in making adjustments, giving up little things that are of no special consequence before being asked to give them up.

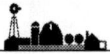

Sometimes friction grows out of differences in religion, for example, church preferences, or over different ways of apprehending or realizing our relations to the Father of all; and these are likely to grow out of radical differences in the make-up. One apprehends the divine through the emotions, another through the reason, another through the imagination, which, after all, are but different forms or phases of faith which brings us into fellowship with the divine.

If, however, they are truly religious in a broad sense, and realize that in the great fundamentals there is little difference between the Christian denominations, there can readily be adjustment that will avoid friction, or, if nothing more, recognition of the sincerity of the other, and an agreement to differ without friction or with the minimum amount of it. The time to adjust these differences is during the courting period. If the religious differences seem to be so great, or the temperament is such that adjustment can not be made, and there is friction, better suspend courting operations indefinitely.

Sooner or later, if we live with one we really love, we will make the necessary adjustments, and all the sooner if children come to bless the home. What a blessed home is that when the pink morsel of humanity in which both have a share, comes into it, looking into the faces of its parents with those solemn, wondering eyes that belong to childhood. What a new rush of love comes into the father's heart for the suffering mother, and what joy into her heart now that she has her reward. Neither of them is a great while discovering that the baby has a will of his own, and if allowed to, will, from his cradle throne, rule the home. Let us hope that in due time another little stranger will install herself in the family cradle. She, too, will have a will of her own, and there will be friction between these two, and all that follow.

I have always regarded children as the all-important crop on the farm, the crop for which all other crops are grown; and the success of the parents in making adjustments to avoid friction, or, rather, in teaching children to avoid friction by making their own adjustments, measures the success in the really big work of life.

All this requires firmness, gentleness, and, above all, genuine affection. We can not deceive children. They read us more truly than we ever read books. Our neighbors may not know us except as we want them to know us, but our children do, for they know us as we are in the home life. The only way to make them believe we are really good people is to be the sort of people we want them to think we are.

Parental folly can develop friction between children by showing a preference for one over the other. Parents are often unjustly accused of partiality by their children. Each child, as it grows up, develops differently from the rest, and has to be dealt with in a little different way. When we look into the matter closely, we ourselves appear different and are different to different people, both inside and outside the home, because each one we meet draws out in us that which is most congenial to him. Often it is not favoritism, but the fact that each child needs different treatment, because it has a different disposition from the rest; hence must be dealt with differently.

Again, the parents themselves may seem different to the different children. To the obedient child, the parent seems and is one thing; to the perverse and disobedient, quite another. This is also true with regard to our relationship with the Divine. David recognized it when he wrote:

With the merciful thou wilt show thyself merciful;
With the perfect man thou wilt show thyself perfect;
With the pure thou wilt show thyself pure;
And with the perverse thou wilt show thyself froward.

It is a sad thing when brothers and sisters do not get along with each other because of friction. Unless great wisdom is shown, it may reach a point where it may blight their future lives. It is an even worse thing when friction reaches a point between father and son, or daughter and mother, so that they lose to a certain extent confidence in each other, and never really get together until great sorrow or disaster bends the stubborn will, and gives them a new vision of life.

The only way to prevent these deplorable consequences is wise, just, and affectionate parental control or government. Children have a keen sense of justice, and openly and sometimes fiercely resent injustice, but they will be forgiven and even make excuses for injustice, if they feel sure the parent loves them, for in the home, as everywhere else, love is the great, big thing. Our influence in the world, and our ability to mold it even in such a small degree as one person can, is measured by our own folks in the home, the folks in our community, the folks in our own country, the folks in the big world. For the love of folks or humanity is the measure of our love for God, and the reflection of our recognition of His love for us.

Uncle Henry

My Dear Folks:

Ever since I was a small boy, I have heard farm women sighing and saying: "Man works to set of sun; but woman's work is never done."

And there is truth in this old rhyme. Woman's work was cut out for her. There must be breakfast every day in the year, and dinner every day, and supper every day except perhaps the Sabbath, when a pitcher of cold milk fresh from the springhouse and a big, fat apple pie, or some doughnuts, used to satisfy the small boy's internal longings, as a cup of tea, or a glass of lemonade with some cake or cookies or pie, or crackers and cheese, seems to solve the problem of now for his grandsire.

Monday was washday; Tuesday was ironing day. Then somewhere in the week was a general clean-up day; a day of patching and darning; a day for baking. After every meal the dishes must be washed and put away. Every day the beds had to be made, the children dressed and made ready for school. There was sweeping to be done, and dusting and scrubbing. There were bruises to doctor, and sore fingers to be tied up. Verily, every woman on the farm had her work cut out for her; and this is true to this day.

But the man, too, has his burden of complaint, only that his work is not cut out for him except in a general way. There is one kind of work in the spring. Nature demands that. There is another kind in the summer, another in the fall, and still another kind in the winter. Most of it has to be done outdoors, and here comes the weather with its imperative "thou shalt not" or "thou mayest." It is time to plant spring grain, but the soil seeps out moisture. It is time to plow, but the soil oozes out moisture. It is time to plant corn, but the soil is too cold. The wheat is ripe, but the skies weep in the daytime, and there are floods at night. It is time to sow winter wheat, but there is no moisture in the ground. When everything has been fair and promising, and there is a prospect of an abundant crop, there may come a hailstorm or a cyclone or a flood, or untimely drought or frost, and cuts everything short.

It is true that part of the farmer's work is cut out for him in a way; some of it he must he must do without fail, or he himself will fail. The livestock must be fed, the cows milked, the horses curried and harnessed, the stable cleaned out. In general, the chores must be done each day. He can plan for that, as women do for their daily work, and do it so regularly that it becomes second nature. The main part of his work, however, must be done out in the open, and subject to all the moods and whims of the weather, of which the only thing that can be certainly said is that it is uncertain.

If the farmer could know in advance what the season would be, and plan accordingly, he would soon have the reputation of a sage. If he knew what the day or the week would bring forth, he could plan his work and follow out that plan as his wife does hers in the house, and as he does his milking and chores. While the woman does most of her work in the house and is thus unaffected by the weather except perhaps on washday, it is at times seriously interfered with by unexpected visitors, sickness or accident in the family or in the neighbor's home. These things can not be helped, any more than the vicissitudes of the weather. We can only make the best of it, whatever comes.

Much of the burden of labor comes from our mental attitude toward it. Complaining about it does no good, and merely saps the strength. We shall have to take it up as we do any other burden of life, as part of life, and do it the easiest and most satisfactory way, which is by putting brains into it.

You can, however, greatly lessen the work in the house by doing away with the unnecessary part of it. For example, where there are no modern conveniences in the home, and where the water has to be brought from a well, have you ever taken the trouble to measure the distance from the well to the kitchen, and multiply that by two, and that by two, and that by the number of trips the wife probably makes in a day, and that by the number of days in a year? You would probably be amazed at the number of miles she has traveled in the year just to bring water to the house. You will then probably seriously consider whether you had not better put in a water system for the house as well as for the livestock.

Is the house so planned that she can do the work which the wife or someone else must do, with the minimum steps, the minimum stooping, and the minimum lifting? Is it not possible to run the washing machine by some kind of power; a gasoline engine or a treadmill driven by dog or sheep power? If the farm is so situated that he can have electricity, why not an electric washing machine, an electric flatiron, and a vacuum sweeper?

I know as well as a man can know how wearing is a woman's work. I know a good deal about the care of children, for I have had a good deal to do with them in my time. But even drudgery is better than no work at all, and no home. No amount of care of children in sickness and in health compares with the desolation of some childless homes. I saw one childless woman the past summer, who,

to put in her time, not only kept her house so immaculately clean that it was uncomfortable, but actually swept the street in front of her house every morning in the year! This, of course, was in town. Work is often wearisome, but the weariness of it is nothing compared to the weariness of the man or woman who has no motive to work for, nothing to do, no one to work for or to love, or who deems life itself a burden.

The farmer can not control the weather; but the very fact that he can not puts him on his mettle as scarcely anything else would do, and particularly so if he is an employer of labor. For this reason, if for no other, farming takes a greater generalship than the management of a store or factory. Merchants and manufacturers deal with things fairly certain; the farmer with things uncertain. He can not tell what the season will be—weather wet or dry, cold or hot; but he does know the average of the seasons, the normal seasons, and he must shape his plans by that. He must become weather-wise, so that during the day, say in haying time, he can shape his plans according to the signs of the weather, not of the moon or the signs of the zodiac. If the weather interferes with haying, he must have work laid out to do to keep his hands employed. If the ground is too wet to plow, he must have planned something else to do.

I used to dread the wet days when I could not work in the field; for I knew to an absolute certainty that my father would have work for me in the barn or in the shop or somewhere else. I did not like that as well as work in the fields. What boy would not rather plow all day than pull weeds out of the garden or would not rather harrow than churn?

Uncle Henry

My Dear Folks:

Thus far in these intimate and, in a certain sense, confidential letters, I have dealt mainly with things pertaining to the home life, the boys and girls in childhood and adolescence or development into manhood and womanhood, the intimacies of home life, its joys and sorrows, its trials and tribulations, the great place the mother holds on the farm, her trials and sacrifices, the health of the farm folks, the work of the farm, and how it differs from the work in shop or office. These are the big things of farm life.

The joy of your life on the farm depends largely on the degree of affection between husband and wife, parents and children, and children for each other. I am not so ignorant of farm life as to suppose there are no differences even in the happiest homes between the man of the house and his better half, no word said that had better be left unsaid, no tiffs or scraps among the children, no feeling on the part of the children of injustice of parents or each other. If every farm home were a little section of heaven fenced off by itself, there would be no need for me to write you these letters, nor would you be interested in any of them. Hence I said "the degree of affection." This love that leads to unselfishness is the big thing in the home life. Unless love reigns in the home between father and mother, between parents and children, and between the children themselves, there will be proportionate trouble and unhappiness ahead.

But while a happy home life is the big thing, it is not the only thing. We need outside fellowship, fellowship of our age and sex. Your better half needs the sympathy of the better half on the next farm. They will understand each other, because they are wives and mothers. The men do not and can not understand them. The boy's happiness is not complete with the fellowship of even the best brothers, even with a dog thrown in. His eyes gleam with new light as he hails a neighbor boy. The girl's life is not complete unless she has another girl to whom she can tell her joys and sorrows, her hopes and expectations. The men are often inconsiderate and thoughtless in this matter. They forget that they are outside more, where they meet with their neighbors in the way of business and in other ways, can compare crops and markets, talk politics, church matters, etc.

Nor does the sound of the human voice over the telephone, nor the handshake at church, satisfy the hunger for human fellowship, though they help. Women like to look into each other's eyes—not Sabbath-day eyes, but everyday eyes. Women all like to talk of their children, their housekeeping, and visit; yes, just visit—a sort of bring-your-sewing-and-spend-the-afternoon visit. This is human nature, and farm folks must have neighbors.

Talk to the wife who has spent a year on a lonesome ranch or even on a lonely road, and you will learn something of the meaning of this natural hunger for fellowship of our own kind and age. Therefore, farm folks must have neighbors; "neighbors," I say—perhaps not exactly intimate friends, or not necessarily so, but neighbors, just neighbors, to whom your heart may warm, and will warm in proportion to the degree of affinity you have for each other. Farm folks are not all alike by any means, and it is safe to say that you don't like and can't like everything you see in each other, but that you need to prevent you from being good neighbors.

Sometimes differences in circumstances and conditions many prevent you from seeing the things in which you are really alike. You may be a tenant on a one-year lease, and may imagine that the landlord's wife or the rich neighbor's wife regards you as a different class, and inferior to herself. In this you may be greatly wronging her. Or you may be the wife of a rich farmer, and may be so snobbish as to overlook the real worth of a struggling neighbor who longs to talk over with you her trials and troubles and joys, who is at heart really a kindred spirit, longs for your fellowship, but fears to approach you; it may be with good reason or may be not. You may be really better than you seem to be to her. You may be bright and intelligent, probably ought to be with the chances you have had. She may be ignorant, her misfortune rather than her fault, but it will pay you to get in sufficiently close touch with her to understand her. With all her ignorance or lack of refinement, she may teach you lessons that you need to learn, and perhaps can not learn from your closest friends. A kind word from you may mean a red-letter day in her toilsome life. Every human being well repays our study, but if we are self-centered and don't like folks, just folks, we may not be able to read aright the story of their lives. I know in this I am asking farm folks to do what most city folks don't do—more's the pity.

I hear some of you saying: my neighbors have bad boys, boys with bad habits, and I don't want to neighbor with them. But your boys will be thrown with them. They go to the same school, possibly the same church—they are thrown together in many ways, and if you are to protect your own boys, you must know these boys and their mother, for you are neighbors. Few boys are so bad that they will not be decent in the presence of a real lady. In their hearts they recognize what is really a higher type of humanity, a human being, especially a woman who likes folks, who is wise, kind, gentle, and sympathetic.

If there is a gang in the neighborhood that is influencing your boy, do not try to break up the gang. Boys naturally run in gangs. So do we older folks. Civilization began in clans, and the clan was to the man what the gang is to the boy. Boys naturally will get together. The way to deal with a gang is not to try to break it up, but to capture the heart of the leader, and the gang will follow.

Take an interest in the sports and recreation of the neighbor boys. They will have them, and ought to have them. There is often quite as much real education in the games and sports of school children as there is in their books. They develop teamwork; when properly conducted, they develop fair play—game for game's sake. It seems to me a matter of prime necessity that farmers in any community get together, make a study of the boys and girls, take an interest in their games and sports and recreation, and so guide and direct them that no harm will happen from the association of young folks together.

Our civilization would advance by leaps and bounds if we could organize in every neighborhood a club made up of the older people, but leaving large room for the younger folks as well. Boys don't dislike the older folks. In their hearts, they are really proud of them, and will yield to the right leadership offered in the right spirit with a good example.

In short, your life is not complete without neighbors, no matter how happy that life may be. You need them; they need you. There was a question asked a young man some two thousand years ago, and by a man who thought he was all right. He asked what was the whole duty of men, and was told that it was to love his God supremely and his neighbor as himself. Then he asked who was his neighbor. He thought his neighbor was the man who belonged to his own class. Then this young man told the story of the man who went down from Jerusalem to Jericho, and fell among thieves, and a stranger came along and bound up his wounds. (Read that story for yourself; you'll find it in your Bible. Then ask yourself whether you are the priest, the Levite or the good Samaritan.) Then he put the question: which of these three proved himself a neighbor to this man who fell among thieves? And the answer was: "He that showed mercy unto him." As I understand this story, it means that my neighbor is not only the man who lives next door to me, or on the next farm, but the man who needs my help. No one needs your help more than the boys in your community. The very first principle of Christianity, which lies at the bottom of all civilization, is that the strong should bear the infirmities of the weak, and should help in proportion to the need of help, of whatever kind may be needed, and the ability to give it.

Uncle Henry

Advice for the Farm Grandparent

My Dear Folks:

When a man reaches the age of sixty or even before that, he is very
likely to spend a good deal of time in reminiscence, that is, thinking
back over the past, the joys, sorrows, trials, and successes of earlier
years. Men have always been that way. I remember when at college
reading in one of Horace's poems a description of an old man, and
was struck with the phrase, "a great admirer of things done when
we was a boy." I confess to spending some time in reminiscence,
that is, when I have nothing else to do, which does not happen very
often, perhaps fortunately.

I am not "reminiscing" now for pleasure, but because it occurred to me that some of my reminiscences might be helpful to
you. I will tell particularly about farm folks whom I have known
who have failed, and my object is to point out the causes of their
failure as a warning to those of you who are young and have most
of your life before you.

I met a woman the other day, who was evidently in poor circumstances, in great need of something to do to make her living,
but of that high-strung type that would scorn to receive a donation
from anybody. I knew her father. He was a farmer owning a large
acreage of the very best land in the corn belt. He was not satisfied
even with superintending his farm and collecting the results. He
was not satisfied with the ordinary way of farming, or in going at
the extraordinary in a scientific way, or even in a plain, common-
sense way. He wanted to do things on a large scale, to bore with
a big auger. I remember that he went into sheep in a large way,
without knowing anything about sheep, or taking time to learn.
His sheep took the scab and foot rot, and his fences were lined
with carcasses. Then he went into fast horses. Finally he got into
politics, and lost everything he had. The last I knew of him before
he died was that he had a room in a shack in town, costing him a
dollar a month rent, eating at a ten-cent lunch counter. What was
the trouble? He wanted to get rich too fast, and to do new things in
a big way. He is a type of a class of farmers who fail.

I knew another farmer, a splendid fellow, owning one of the
finest tracts of land in the corn belt. He, too, wanted to do things in
a big way. He became embarrassed and sold his farm and bought a
much larger tract in a new country. On his first farm he had engaged

in a business he knew nothing about. On the second farm he tried another new business, and lost everything. He died not merely bankrupt, but a pauper, and was buried by his friends.

I knew another farmer, farm-born and bred, who moved to town, engaged in business, made money rapidly, and went back to farming again. I know a good deal about his operations. He bought some of the best land in the corn belt at about $15 an acre, broke it up, and rented it out. He did this on such a magnificent scale that, to my own knowledge, one year his income from rents was $30,000. If he had simply held onto that land and looked after his rents and improvements, which he understood as well as any man I ever knew, he would have been worth millions. But he was not satisfied. He engaged in business in the farther West, business that he knew nothing about, and the man who was supposed to be worth a million eventually had nothing left but his house, and I fear even that was mortgaged.

I mention these cases because they are typical of many others. Not one of these men, and they were all personal friends of mind, had a single bad habit so far as I know. Not one of them was even suspected of drinking, nor of dishonesty. All were highly respected. And yet, with good opportunities for success, they failed simply because they were so eager to get wealthy fast, and overreached themselves. Many who looked up to, admired, and flattered them because they were prosperous turned from them as soon as their credit was shaken. The heartlessness with which people who have received favors will turn down their benefactors in the day of adversity is simply appalling.

I have known other farmers who failed, and failed in a worse sense than these. They died rich, "cut up well." Their failure was not in the loss of fortune, but in losing the confidence and respect of their children; and the grass had not yet grown over their graves before the children whom they had failed to rear properly were fighting each other over the spoils, disgracing themselves, and disgracing the memory of their father with their eagerness for more than their fair share of the estate. You need not be very old to know cases of this kind. Some of them, when too late, saw the terrific mistake they made, and tried to remedy it by their wills, but these wills were broken, or the estate squandered in trying to break them, and they and their families were soon forgotten. There are few worse failures than this.

You have all known of farmers, perhaps in your own neighborhood, who did well as long as they confined themselves to the lines

of farming to which they were accustomed, but become dissatisfied with the slow way of making money, and went to speculating on the Board of Trade. They perhaps made money the first turn, and said: What's the use of working my life out on the farm? All I have to do is to study the crops, and make use of my superior knowledge of buying and selling on the Board of Trade.

I have never known a farmer who made a permanent success on the Board of Trade. He may think that he knows all about crops and prospects, but at last wakes up to the fact that the men who deal on the Board of Trade are familiar with the crop prospects all over the world, and have sources of information to which he has no access. And even these last, with all their shrewdness and wisdom, seldom die rich.

If the farmer wishes to be worth a million, he is not likely to make it farming. Here and there one does it by the advance in the price of land. The unearned increment alone makes them rich. But any one of you who owns a farm can make a good living, can rear your children to habits of industry, economy and thrift, can teach them to deal right with their neighbors and friends and business associates, can die in peace, your money remain with your children, and your memory be honored. But if the get-rich-quick microbe gets hold of you and you take to speculating, the chances that you will die worth a million are very slim indeed, and the chances that you children and neighbors will honor you are still slimmer. If you want to know whether it is worthwhile to die worth a million, just make a study of the sons and daughters of men who have made a million or more. I know of nothing that will get the get-rich-quick microbe out of your system more effectively than that kind of study. The chief end of man is not to acquire a lot of money or property, which he can not possibly take away with him, and the future use of which he can not regulate, but to rear a family of sons and daughters who will make life in the community better worth living. If he succeeds in that, he has something that money can't buy. If he fails in that, then, no matter how much he leaves, he must be put in the class of farmers I have been writing about in this letter.

Uncle Henry

My Dear Folks:

If you ever asked a bright little girl, just after her sixth birthday: "How old are you?" did you notice the sparkle in her eyes as she answered: "Going on seven!" If you ask her again five years afterward, she will say: "Going on twelve!" She is growing older, and is glad of it. But if you ask her the day after her twenty-fifth birthday, she will not say: "Going on twenty-six," and there will be no particular sparkle in her eye unless she is about to be married. In that case the sparkle will be there all right, and plenty of it. In any event, she is more likely than not to give you an intimation, more or less polite, that is none of your business how old she is. Even if a ring sparkles on the third finger of her left hand, she is pretty apt to give you an evasive answer.

If you ask a boy past twelve how old he is, he will say: "Going on thirteen!" and until he is past twenty-one he is still "going on." After that time it depends. If by the time he is thirty-five he has become his own employer, or has an assured position, he won't care; but if he is looking for a job, he will let you guess at his age, and may not admit the correctness of your guess. Up to a certain point, varying with circumstances and conditions, we are all anxious to grow older, rejoicing in it, glorying in it. The girl looks forward with unspeakable gladness to the time when she will wear long dresses, do up her hair, and have a beau, like "big sister." The boy looks forward eagerly to the day when he will be of age, have a vote, and become his own master—so he thinks.

We older folks like to see these young folks growing up. So did the Psalmist when he said: "When our sons shall be as plants grown up in their youth, and our daughters as cornerstones hewed after the fashion of a palace." We like to see the rosy flush on the cheek of the lovely young girl, the lithe, supple form, the firm, round muscles, the fawn-like movements, the abounding spirits, and, perhaps, the lovelight in her eyes. We like to see the broad shoulders and sturdy limbs of a young man, the budding mustache, the smooth-shaven cheeks, the growing hardness of the muscles, the keen, steady, honest eye, the open, serious face. All this delights us, not so much for what we see in these glorious young people now, but for its promise for the future. They remind us that we were once young, and bring back recollections that will never be entirely erased from the tablets of memory.

But there comes a time when for most of us the vision changes. If the girl has married and is blessed with the crown of motherhood, she does not grieve so much over the passing years. She lives for her husband and children. She has now a strong arm to lean upon and in her children she lives over her girlhood, with its joys and sorrows, its hopes and fears. She does not then care so much if crow's feet gather about her eyes, if she loses her willowy form, if her hair—the glory of her girlhood days—becomes thin, and she must piece it out with a switch.

If her marriage has been a mistake, and especially if she has not been blessed with the care of children, whether her own or those of another who need her care, she is likely to look on the coming years with foreboding. But whether the forebodings will be realized or not depends mainly on her attitude toward life. It is folly to mourn over either the mistakes or misfortunes or disappointments of the past. There is a beauty of character which dims all the radiance of beauty of face and form or outward adornment; and this beauty comes of its own accord to those who, whether married or unmarried, take up resolutely and patiently the duties of each day as it comes. We have the best of authority for saying that even "the hoary head is a crown of glory, if found in the way of righteousness." Every age has its own type of loveliness, and we older folks become ridiculous only when we attempt to conceal what was once ours, but is now the heritage for a time, and for a time only, of the following generation. Beauty of character comes only to those who have learned to live for others; nor can paint or powder or dress erase or long conceal the lines which a selfish life has graven on the human countenance.

Someone may ask: Why are you telling us about the future? Why don't you let us enjoy the present and live while we live, without pointing out to us what may come afterwards?

Because I feel quite sure that if I can get your attention fixed on two or three things, I can help at least some of you to a sweeter and happier life.

In the first place, you can't help growing old. You are older now than when you began to read this letter. You will be still older when you finish it, unless you throw it down in disgust—but that won't stop your growing old. You have started on this mysterious journey that we call life, and you can't stop or go back. If you could, and let the rest go on, you soon would be the most miserable of mortals. You must go on.

In the second place, growing old is not half as sad as you think it is. In fact, it is not sad at all, unless you make it so. Every stage of life has its own particular joys, as well as its own particular sorrows and trials. Both are necessary to a proper development of character; and we must pass through them at the right age or miss the blessing of them. Dolls, school, sweethearts, lovers, husbands, children, grandchildren, and the literature pertaining thereto—this is a natural order for the girls. Games, marbles, movies, school, a settled business or trade, best girls, wife, children, grandchildren, with the literature thereto belonging—this is the natural order for the boys. Each brings a joy and an education peculiarly its own.

The latter stages of life are the better, the fuller, the richer. The man or woman who lives right and plays the game fairly all the way through, does not realize that he or she is getting old. They are too busy to think about it, and get too much out of life to dread it. When I was a boy, and aunts or uncles came to visit us, and I was told how old they were (none of them over sixty), I thought it must be awful to grow old. I looked at their gray hairs or bald heads, at the men carrying canes and the women wearing caps, and said to myself: "That's awful; I hope I'll never look like that." But I was fifty before I knew it, and I did not feel old; then sixty, and I was still too busy to notice it; then seventy, and still I felt young in spirit. If one keeps busy and interested in things, I doubt whether he will ever feel old.

The third thing I want to get fixed in your heads is that if you are to get the real fun, the joyous music of life, you must grow old gracefully; and the time to begin that is right now. Live sanely and cleanly, take care of your health, learn how to feed yourself, keep clean inside as well as out, trust your Heavenly Father, and show it by your dealings with your fellow men.

Your life will not be all smooth. You will have grievous disappointments. The future will look dark at times, and you may perhaps feel like complaining bitterly that your life is hedged in; but my own experience is that many of these disappointments are among the best things that have happened to me. You must go through trials and sorrows, often deep waters, if you are to develop the sweetness and patience, the endurance and broad charity, the pity and compassion for others, that go to make up a life of power and real influence. I remember when I was a boy reading something like this from one of our New England poets:

Hearts, like apples, are hard and sour;
The coming years will mellow them.

Some of us need a good deal of "sugaring off." Some of you may not know just what that means. When I was a boy and helped to make maple sugar, I noticed that it had to be boiled and boiled and

boiled. Then came a time when the heat was lessened. We called it "sugaring off"; and sugar never tasted so sweet as when it was just sugaring off. Nature does the same with the corn fields. We have the fierce heat of summer; but when September comes, the heat lessens, and the ears of corn are, so to speak, gradually "sugared off" to full ripeness.

Let me beg of you not to look forward with foreboding to the future. The good Lord has a plan in your lives—every one of you; and if you will trust Him and do your part, you may not die rich or honored, but you will be loved, for you will develop graces of character which will compel the affection and confidence of those who know you.

Uncle Henry

My Dear Folks:

Farmers, like all other folks, will grow old. Once started on this mysterious journey called life, we can not stop until we are stopped by death, through disease or accident, or old age, and the longer we live, the older we get in years, if not in spirit.

We usually have a right good time in the beginning, nothing to do but sleep and get outside of a well-balanced ration, with a fore-taste of future trouble in the shape of colic or misplaced pins. Then comes childhood, in which we get in touch with a world altogether new and wonderful. Then comes youth, for the most part joyous and carefree, but not without its grievous trials and sorrows, at which we laugh in after years. Then comes manhood and woman-hood, the founding of a home, with all that it involves for good or ill. Then the weary round, the struggle to get a good position and maintain it, to get and keep our "place in the sun," to rear a family of children that will do honor to themselves and to us. For thirty years this is the main work of the grown-up farm folks. For most of us they are the happy years. Nonetheless, there is plenty of grief scattered through these busy years; hard work, any amount of it, sometimes loss of crops, sickness among the livestock or, worse still, sickness in the home and the death of loved ones. But all these things, if met bravely and borne patiently, are but steps in the build-ing of the structure, the rugged character which is typical of the life of farm folks.

When we are in our prime, and full of life and vigor, with our children about us, and we feel that their future is in our hands, we don't mind hard work. We rather glory in it. We face adversity with a stout heart. We are happy because we are able to overcome the difficulties and dangers.

There comes a time, however, when we begin to realize that we are not what we once were. We feel that we have passed over the top of the hill, the maximum of our energy, physical or mental, or both. In other words, we are beginning to get old. Some men get old physically in middle life. Others are comparatively young when their neighbors think they ought to be getting old. Others, by reason of sickness or accident, are physically old before their time, or because they are mentally old prematurely, in other words, mentally lazy. I have seen some men that I felt were born old.

There is the same and even greater difference in mental aging. A man is mentally dead when he ceases to get new ideas, when he fails to react to the impressions that reach him from the world around him, from men and things. I have known men, and women, too, who reached this point before they were thirty, persons whose self-conceit was so colossal that they had in their own minds boxed the compass of the intellectual world and knew all there was to know. Thereafter they were known as "dead ones."

On the other hand, I have known men and women who were mentally young, keenly interested in all that was going on around them, readjusting themselves to the changing world at eighty, and even eighty-five—old in years, but young in spirit. A preacher of ninety-two called on me one day for suggestions on poultry-keeping, remarking that it was only in the last year or two that he had realized how much pleasure and profit there was in keeping chickens. Last Sabbath a preacher aged ninety-eight walked some distance to church to deliver one more message for the master.

In the natural order of things, physical aging comes before mental for very obvious reasons. The soul is immortal. The body is only the casing, the vestment with which it is clothed for this phase of its existence, and through which it can express itself, changing it from day to day, and year to year, and at last, laying it aside as a garment no longer needed. The wearing qualities for this casing are determined by the care taken of it and by the vigor, "ginger," "pep" of the undying mind. But no matter how much "ginger" a man has, this body will in time begin to show signs of wearing out. In other words, it begins to get old; and the spirit within should take note of that fact and not ask too much of it.

When a boy, you jumped out of bed in the morning, jumped into your clothes and ran downstairs, eager for the day's work and fun, perhaps more eager for the fun than the work. As you grew older, you were more leisurely about jumping out of bed and dressing, and walked downstairs, planning your work for the day.

There comes a time, however, when you are in full sympathy with the song that Harry Lauder sings:

It's nice to get up in the morning,
But it's nicer to lie in bed!

You feel like you want to stretch and yawn, perhaps take another nap. You feel that your knees are getting a little stiff and need to be limbered up a bit before you put them to work. Even after they are limbered up, you may like to lie abed a little longer. It does not bother you if breakfast is a little late; but it does provoke you to think that you are not quite the man you used to be. You are quite as anxious to get the work done as you ever were. Your head is quite as clear as it ever was. "The spirit is willing, but the flesh is weak." You try to lay out the work in such a way as to give you an easy job, mostly bossing. You take rather kindly to chores. In harvest you prefer to ride on the mower or the binder. In haying time you prefer to be on the wagon, if the hay is pitched up to you. If there is a loader on the job, you would rather drive the team.

And so you go on for a few years, until your son marries some nice girl, or your daughter marries a nice young fellow, and you get a hint that they would like to rent the farm. Perhaps you may not exactly like it, for it looks a little as though they were trying to shove you off the place. Perhaps you propose that they move in with you, and the son, or son-in-law, take the short and heavy end of the doubletree, do all the work and let you look after the chores and odd jobs. You notice a shadow on your wife's face about this time; and before long it will dawn on you that no house is big enough for two families; that two queens, even mother and daughter, can not be quite happy in the same hive.

Then you begin to think about moving to town. If the good wife should take kindly to that suggestion, it would not be surprising. She has had rather a hard time of it all these years. You could get help on the farm, but she could get none in the house, and the more help you had, the more work it made for her. She would naturally feel that a good long rest would be a nice thing for her, and suggest that the rent of a quarter section should easily keep you in comfort in town. You think so, too. Don't be too sure about that. Living out of the grocery store and meat market is a far different matter from living out of the garden, the poultry yard, and the barnyard.

Otherwise she can get along in town, for she has her household duties to keep her busy; she has her church and its minor societies. But you will find that you are not so tired as you thought you were, Mr. Man-getting-lazy. After you get your house fixed, the garden made, and the yard fixed up, you will want something to do; but alas, there are enough old fellows in town already to do all the old men's work.

You will not take nearly so kindly to town folks as you thought you would from your meetings with them before you moved to town. In fact, they may not take very kindly to you. They will not welcome you to their stores if you have nothing to sell or buy, nor to their offices, if you have no business to transact. You will lose the political influence you had on the farm. Your tenant will in all probability be a bigger man politically than you are, simply because he lives on the farm.

You will find yourself longing for Saturday to come, when you expect your old neighbors to come to town to trade, and you can get in momentary touch with the old life. If you have rented for cash, you will probably quit thinking about farming, will read local news and telegraphic dispatches instead of editorials. This will age you mightily, and that will react on your health. You may even fail to renew for *Wallaces' Farmer* on the advice of your neighbor, who may tell you that if you keep on taking it, it will make you homesick for the farm. In fact, men have told me that they have received just that advice from other retired farmers.

So you will do well if you think over carefully this matter of moving to town. I don't say you should not. There are circumstances under which it would be well to do so. In case you should drop off suddenly, it is quite likely that your widow would be better off in town; but you are not likely to drop off nearly so soon if you stay on the farm. Before deciding to move to town, however, I would suggest that you consider the alternative. It is no new advice I am giving you. I have spoken of this often. I speak of it with more confidence, because men and women who have taken my advice in this have written me, and sometimes come in to tell me how happy they were that they did not move to town.

What is that alternative? Rent your farm to the best man that you can find, a man who will take care of it and maintain fertility, a man whose character is such that you can reasonably expect him to stay on it as long as you live, no matter what the time of the lease you give. Build you a house of your own on say ten acres, or five, a house suitable for yourself and wife and what unmarried children are at home. Don't build it too big, but be sure and have room in it for your visiting children or grandchildren. Keep a cow; a horse or two, if you like to drive. If you are able to afford an automobile, get one in addition to the horse or two. Help your wife to grow some fine chickens. Grow improved seed corn or some new variety of grain. Keep in touch with your old neighbors. They will be nearer and dearer to you as they grow old along with you. Keep in touch with young life. Take an interest in all farming operations. Keep in touch with your church, your Grange, or farmers' club. This, I think, is the best way of rounding out a farmer's life.

It may seem strange to you, but I never noticed in Great Britain or Ireland any indication of farmers wanting to move to town. They stay on the farm. Even landlords stay on the land. They keep up their interest in the farm life. We will get to that point in this country by and by. We will not look up on the farm as something to work out, to skim or mine, and then run away from, but as a home, a homestead that will remain in the family for generations to come. We will begin to see before long that country life might be the best life, which it will never be so long as farmers and their families run away from it as soon as they can.

Therefore I suggest that you think over very carefully this alternative of moving to town. You will find, if you once move to town, that it is not so easy to get back.

Uncle Henry

Part Three

Uncle Henry's Own Story

Letters to the Great-Grandchildren

Henry Wallace, in His Study, 1914, Dictating From Personal Memoranda the Letters to His Great-grand Children.

The Wallace Boys' Preface to *Uncle Henry's Own Story*

These letters, written by Henry Wallace, and addressed to his great-grandchildren, were the result of a chance suggestion made some years ago. Mr. Wallace had lived a very full and eventful life during a most wonderful period of the world's history. Between his boyhood and old age, a transformation had been wrought in methods of living and in civilization itself. Means of transportation by land, sea, and air had been wholly changed. The world had emerged from a period of hand labor to machine labor, with a revolution in the lives of laboring people. It was a period of invention, discovery and wonderful progress in transportation and science; a period of world-wide evolution in agriculture. All of this he had seen, and in some of it he had played a very important part.

It was suggested to him that the ordinary biography or even autobiography fails to tell the things that people most like to learn about. That his great-grandchildren, for example, would be intensely interested in the sort of life he lived as a boy and a young man. They would like to know about the things in which he had an active part and in which he was vitally interested. They would like to know of the manners and customs of the people with whom he drew and lived. Why not, as he had leisure, write a series of intimate letters to young folks, who probably would be coming on years afterwards—the sort of letters that would reveal his own personality as no biographer could do it?

The suggestion was received with instant favor, and very shortly afterwards the first of these letters was written. From that beginning, during the next three or four years, as the spirit moved him, he wrote additional letters, the last one but a few months before his death.

We have felt that the thousands of people who admired and loved Henry Wallace have a very real claim to share these letters with the great-grandchildren to whom they were addressed: and we began their publication in *Wallaces' Farmer* in the autumn of 1916.[140]

Wallace Publishing Company
Des Moines, Iowa

My Dear Great-Grandchildren:

At this writing, none of you have put in an appearance as yet, and probably will not for some years to come. Nevertheless, I am morally certain that you will appear in due time.[141] You will make your appearance in a world so different from that in which I made my appearance, some seventy-five years ago, that when you read my description of my world, you will no doubt wonder how I managed to get through. You are coming into a world that has railroads and street cars and telephones and telegraphs and automobiles and flying machines and Sunday papers. You have electric lights and gas, bathrooms and sewage and furnace heat of various kinds, pianos and piano players, and rugs, to say nothing of electric carpet cleaners. You have baby carriages that fold up, dolls that can talk, washing machines and sewing machines run by electricity. Your ironing is done by an electric iron, and, for all I know, you may be having all your cooking done by electricity.

When I was born, we had none of those things; at least there were none in our neighborhood. They had railroads of a very primitive sort "down east" and also steamboats as primitive. I never saw a railroad till I was twelve years of age; never rode on a railroad train till I was eighteen. You will think that your great-great-grandfather and great-great-grandmother lived in a very primitive way. So they did; but they lived happily and reared a large family of children, none of whom except myself, however, lived to be thirty.

I am writing these letters for your information, that you may know these matters in detail; not for information solely, but because I wish you to realize that you would not have had the comforts you have, and the opportunities, educational and otherwise, that you now enjoy, unless the people who lived in my day had faced the difficulties and endured successfully the trials and hardships of that day; and it is important for you to know the steps by which the world has made progress, giving you the advantages and opportunities which you now enjoy.

The progress of civilization has been slow but fairly steady. Each generation is apt to look back upon the past one as slow, old-fogyish and out of date, forgetting that they are indebted to these seemingly slow-going people for the privileges they themselves enjoy. It will not do for one generation to put on airs and imagine that they are the only people, and that wisdom will die with them. We owe a great deal to our fathers and mothers, our grandfathers and

grandmothers. You are enjoying luxuries which kings and queens, with all their wealth and power, could not possibly have secured two hundred years ago.

I want you to see how civilization has developed slowly but surely, step by step, through toil, privation, struggles, victory sometimes, and again partial defeat, but, on the whole, making a gradual advance. I wish you to realize also that with all their disadvantages, people were just about as happy in those early days as you are now, or ever will be.[142]

Your great-grandfather,
Henry Wallace

My Dear Great-Grandchildren:

In my boyhood days, amusements and recreations were somewhat limited—perhaps not more so, however, than in many rural communities today.[143] There were some, however, that I am quite sure the boys and girls do not have now. For example, we had husking bees. The farmer who intended to give entertainment to the young folks of his neighborhood by having a husking bee snapped his corn, put it in a long pile, perhaps three feet high, three or four feet wide at the bottom, and coming to a point at the top. Then he invited in the boys. They chose two captains, who chose sides. They then divided the pile evenly, and the question was which side would get through first. With the husking bee, there was usually something doing at the house, perhaps an apple paring or a quilting. After the work was done, there were "eatin's," and after that usually some dancing of the old-fashioned sort, and, of course, a fiddler. My father and mother regarded dancing as something which belonged to the unregenerate, and I never learned to dance. Possibly I was all the better for never learning.

The boys had more forms of amusement than girls. One of the favorite sports in the fall of the year was coon hunting. If there was a pack of hounds, or even one or two in the neighborhood, there was some real sport in hunting coons on a moonlit night, listening until the lead hound struck the trail, then waiting until the barking showed that they had treed the coon. Then came the interesting problem how to get down. It must be done either by climbing the tree—which was sometimes done—or by cutting it down. This we

did if it was not too near the house, for some of the good old farmers regarded the cutting down of a coon tree or a bee tree as something verging on sacrilege. If the farmer was particularly irritable on this point, of course this was a good reason why we should irritate him. Such is the perversity of human nature. When the corn was in the roasting ear, it gave zest to the sport to "hook" some roasting ears from the field, build a log fire, and roast them while the hounds were trailing another coon.

There was a good deal of recreation in the old-fashioned singing school. The teacher usually had three or four schools which he conducted on different evenings of the week. This gave a chance for the boys and girls to get together. I do not know how much singing we learned; but I do know that about the only tunes I can sing to this day are those that I learned in the old singing school. One of the interesting things about singing school was taking the girls home; for of course they must not be permitted to go through the wood alone. The boys usually got out first, and waited outside the door. Then when the girls came out this was heard: "May I see you home?" or "Will you accept my company?" Sometimes she gladly said yes, sometimes no, in which case he was said to have "got the mitten," and great was the glee of the other boys who overheard it, and much he had to endure afterwards.

There was one diversion in our neighborhood that we greatly enjoyed, and it was called "swabbing the river." The Youghiogheny is a rather narrow stream, with riffles where some harder and more enduring rocks come to the surface every half mile or so, and between these riffles stretches of water several feet deep, and sometimes with deep and dangerous holes. We could "swab" the river only when it was low, in August. Farmers would say to their boys: "Now, if you will get the manure hauled out, you may have a day's fishing." The boys cut grapevines and brush and made a rope about as thick as the body of a small horse, and stretched it out across the river. In the meantime, another detachment had thrown up dams in a riffle leading into a pot. After the fish were scared into that pot by the swab, which frequently had rye straw fastened to its underside, they must themselves catch the fish by hand or with a net. Sometimes we caught many fish, at other times none or a very small number; but, fish or no fish, we had a fine day's sport, and had fine appetites for supper.

Where there was a sugar camp, or a sugar bush, as it is called, we had a good time during the month of February, when it came to "sugaring off," in which the boys and girls could take part. In spite

of these things, as I look back over those days, I realize that the amount of recreation and amusement was pitifully small.

My Dear Great-Grandchildren:

It may interest you youngsters to know something about what kind of barn your great-great-great grandfather built, a few years after he married and settled down and began to get a bit ahead in the world. Fashions change. The hat or coat worn last year will be out of style this year. Fashions in barns change also, but there is much more sensible reason for the change of fashion in barns than in clothes. Barns are built in every country, in every age, to suit conditions, and the student of agriculture could, by studying a barn in any country, and in any age, however remote, make, from that study alone, a pretty accurate guess as to the agriculture of that country, just as a scientist, by finding a bone of some extinct animal, can piece out the rest of it and tell us what kind of an animal it was, what it lived on, and when. There was an old barn on the place when my father bought it—built of logs, and made to meet very primitive conditions. I remember seeing it once. In fact, it is one of the first things I can remember.

I have no distinct recollection of the building of the new barn; could not be expected to have, as I was only about three years old. It was built in 1839, the year my oldest brother was born. The foundation was 30 x 70, but the long beams projected out nine feet over the width of the barn, making it what was called an overshot, a place where cattle could get in and out of the wet when it was not desirable to put them in the stable. It was of the bank type—that is, the east and north sides were partially underground, the bank being cut away from one to four feet. The object, I suppose, was to pro- vide warmer quarters for the cattle in winter. The barn proper was divided the short way into four sections, a mow on the east and on the west, with loose boards for the floor, and two threshing floors in the center. The walls of the lower story or basement were of rock; and the entrance to the threshing floors was by a bank and a bridge.

This barn was built to last. It has now stood over seventy years, and, with the excepting of needing re-shingling every twenty-five or thirty years, is apparently good as ever. Roofs lasted longer then than now, for the reason that you could get better shingles, and for the further reason that they used iron nails instead of the miserable steel nails that rust out in a few years. Timber must have been cheap and plentiful when that barn was built, judging from the size of the great hewn logs used for beams and floor sills, else sills would have been placed on edge rather than on the flat side. There was good enough oak and sugar-tree timber in it to build two or three barns of like capacity.

You may wonder why my father built a large and expensive barn like this several years before he built a new house. Here is another illustration of the changes in fashion. My father deemed it necessary—and in this he did not differ from his thrifty Scotch-Irish and Pennsylvania Dutch neighbors—to build a large barn in order to save the grain crop from damage and to shelter the livestock, and thus lay the foundation of future prosperity. The new house could wait until afterwards, for in those days people were used to self-denial. Moreover, the size of the barn had something to do with the farmer's standing in the community. When a farmer in the next township built a barn a hundred feet long, some of the old fellows whose barns were seventy feet long, thought that was going a little too far, and rather suspected it was not all paid for, and the more so because he had for a weather vane an enormous cow.

These great barns, when full of hay and grain going through the sweat, were sometimes struck by lightning, and the first insurance company I ever heard of was a verbal agreement made between my father and a number of his neighbors to rebuild the barn of any one of them in case it was destroyed by lightning. None of them were ever struck, and hence the soundness of this novel insurance company was never tested, but I have not the slightest doubt that the agreement would have been carried out. A man's word pledged to a neighbor in those days was a sacred thing, whether given as a pledge for the payment of money, for the fulfillment of an agreement or contract, or anything else.

Some of you youngsters, who are accustomed to seeing hay barns with a great door let down near the top through which the hay is lifted by horse power or gasoline engines, may wonder how we got the hay and the grain into this barn, and what need there was of two threshing floors. You probably do not know what a threshing floor is. The place to do the threshing, of course. This was done in two ways, with a machine, or in the old scriptural way, described in Isaiah, 28:27, except that in Isaiah's day the threshing was done by oxen, as in days of Moses. You remember, Moses said, "Thou shalt not muzzle the ox that treadeth out the corn."

The machine was a very simple affair, an enclosed cylinder and

concave driven by horse power, sometimes four horses, sometimes six, sometimes eight—the power being transmitted from the horse power outside to the machine inside by a tumbling rod. The sheaves were thrown out of the mow into a table by the machine, and there unbound by hand, in order that the speed might not get checked by a knot passing through the machine. They were fed into the machine, which was so located that the straw and the grain were shot into the corner.

The hardest work would fall upon the man who stood at the tail of the machine and raked the straw away in a direction at right angles to the machine. There were about five or six shakers, with forks, who passed the straw from one to another, shaking it as they passed it out, until it reached the stackers on the outside of the barn. After running about three or four hundred sheaves thru the machine, it was necessary to stop and "cave up," that is, shove the grain that had accumulated to the far corner on the opposite side of the threshing floor.

Threshing was regarded as hard and dusty work. As the work was done cooperatively, neighbor helped each other in turn. After I was fifteen, I had a good deal to do in "paying back," as my father choked up with the dust, and I went in his place. I rather liked it, because there was plenty of company (there was more isolation in farm life then than now.), but mainly because the eating on those occasions was the best. Nothing was regarded as too good for threshers. What fine eatings there were, and I suspect it tasted better in those days, because I never lacked an appetite. And then how handsome the girls looked as they waited on the table.

But why two threshing floors? Because the size of the barn necessitated it. My grandfather had a barn of the same type, but shorter; he had but one threshing floor in it, for the reason that he could pitch from either mow into that threshing floor. This could not be done in a barn seventy feet long. The mows were at either end, each seventeen feet wide, and, to save time in pitching, each required a threshing floor adjacent, wide enough to store a day's threshing of grain.

The other way of threshing was tramping it out with horses. This method was not used with grain except to supply the family with flour until threshing could be done. For you must understand that in my early days we did not buy flour at the store, by the sack, or barrel. When we wanted flour, a three-bushel bag was filled with wheat, thrown on the horse's back, a boy put on top, sent to the mill a mile away, and told to wait until the grist of grain was ground, and bring home the flour, bran, and shorts, less the miller's toll—one-tenth. He was told he might go fishing while he waited, if he liked. This was my job for some years, and I remember that the three-bushel sack lasted us about ten days. That was some years after we built the new house, however, and had plenty of company. As it required a man to put three bushels of wheat (180 pounds) on the back of a rather large horse, if the boy let the bag fall off, as I one time did, all he could do was to wait until someone came along, put the bag on the horse, and lifted him on top of it.

Corn was nearly always tramped out, the threshing floor being covered to a depth of about a foot. The horses went around and around on the floor, a small boy usually claiming the privilege of riding, an attendant shoveling the corn out of the center into the track of the horses, and shoveling back into the center the shelled and corn and cobs.

To get the hay and grain into the barn, the loaded wagons were driven into one or the other of the threshing floors, and the load pitched by hand into the mows on either side. Great beams were placed crosswise of the threshing floor, about the height of the barn, which was about the height of the loaded wagon. On these beams, platforms were made, the hay or grain was pitched onto these platforms from the wagon, again pitched into the mow, and then, by another hand, pitched back, with one or two boys to tramp— hot work on a hot day. After the mows were filled, one threshing floor was filled, and afterward the space over the other—these last mainly with grain, preferably with oats.

If you are interested in farm matters, which I hope you are, you may want to know what kind of corncrib we had. Well, you would have laughed at it. It was four feet wide, eight feet high, and long enough to hold the bulk of the crop. It was located four to six rods from the barn. As I remember it, it held about six hundred bushels. The rest was left in the shock, for we nearly always harvested our entire corn crop, and husked it as need be after hauling the fodder to the barn. Farmers in those days never would have thought of committing the extravagance of wasting the forage. The reason for the narrow crib was the very significant one that corn would not keep in that country in a wide crib.

My Dear Great-Grandchildren:

Though human nature in all generations is essentially the same, manners and customs differ from generation to generation. For example, the love that grows up between young men and maidens in every generation is essentially the same. By some mysterious law of attraction, which even they themselves do not understand, the young man is attracted to the maiden and the maiden to the man. They have the same hopes, delusions, lovers' quarrels, reconciliations—but the manner of expressing them changes somewhat.

In the days of my youth, the young man did not buy a shiny red buggy and go in debt for it when he was courting. Instead, he managed, if possible, to get a good-looking horse, saddle, and bridle. If we younger boys saw any of the older boys taken with a sudden love of horseflesh and the trappings thereof, and saw one of them, in the cool of the summer evening or in the storms of winter, dressed in the best "bib and tucker," going to a certain house along about eight or nine o'clock, we knew there was something going on. If the horse was tied to the hitching post, we knew there was nothing serious yet, but if the young man put his horse in the barn and took off the saddle, then it was evident that he had become a "steady," and something pleasant was imminent. If there was a quilting bee, and certain mysterious nods and winks among the girls, then we were quite sure of it.

When the engagement came, as it generally did in due time, there was no diamond engagement ring, frequently no ring at all, sometimes a plaited gold ring with two hearts held together by a cupid's arrow. There were no prenuptial dinners, no "showers." The wedding was a rather simple affair. Among the more well-to-do, a preacher was called upon to tie the knot. Among those of little financial or social standing, a visit to the squire in the neighboring town was all the wedding there was. There was no license law in those days, nor for long afterwards. If you and your sweetheart were agreed, and a preacher or a squire was handy, a brief ceremony was all there was of it. Among the well-to-do and those of some social standing, there was usually considerable of a wedding—that is, a number of guests, relatives, neighbors and friends were invited, and there was a dinner—the best the house could afford. The bride and groom did not go on a "bridal tower," but usually spent the first evening and night at the bride's home. The next day was the "infair," or the reception at the home of the bridegroom, and it was a matter of pride on the part of his folks to have the dinner as

sumptuous and elaborate as that given by the bride's folks. After that, the young folks settled down in their own home, which usually had been provided and furnished beforehand.

There were elopements then, and separations, just as there have been from the beginning of time, and will be to the end, I suppose. There were some funny advertisements in the paper, however. In case the wife could stand it no longer and "vamoosed," the husband put an advertisement in the county paper, substantially as follows (I quote from memory): "Whereas, my wife, Elizabeth, has left my bed and board without due cause or provocation, I hereby warn all persons from harboring her, as I will pay no bills on her account."

Sickness and death came then as much as now. In our neighborhood, a hired nurse was never heard of. There were always one or more women, married or single—mostly single—who were nurses by instinct, and these were called upon in case of sickness. Every neighbor felt himself bound to lend aid in some time of trouble, for he did not know when he or his might be in like need. Where a night watch was required, there were usually two, one watching till one or two o'clock, and the other from that till daylight. We knew that some diseases were "catching," but we had never heard of bacilli, microbes, or germs. Unfortunately, we did not know much about sanitation, and hence there was more risk in waiting upon the sick than there is under modern conditions in the average home.

We had never heard of embalming the dead, except from the Bible, and no undertaker was called in. If a man died, the body was washed and dressed by some men in the neighborhood; if a woman or child, by some neighboring women. It was then laid out on the "cooling board" and a watch kept. There was always a "wake," that is, a couple of neighbors sat up in the same room or in an adjoining room, with the door open between, with the other doors and all windows closed, for fear of cats, which it was believed were ever eager to disfigure the body of the dead. An undertaker furnished the coffin and hearse, unless the family was in rather poor circumstances, in which case the farmer's spring wagon took the place of the horse. It was considered a very neighborly act to dig the grave for a neighbor, for there were no incorporated cemeteries and no sexton to do it. There was no decoration of the grave, no lining with cotton and evergreens. There was no mechanical arrangement by which the body could be slowly lowered. This was done with lines, generally taken from the farmer's harness. Two at each end lowered the coffin into a rough box, then the grave quite full, smoothing and leveling the surface. It was regarded as rather bad form for any family to

leave until he had seen the last shovelful of earth put upon the grave and smoothed down. Funerals were generally well attended. Some kind neighbor or neighbors stayed at the house to prepare a dinner or supper for the returning mourners with some near friends who were invited to dine with them.

There were no costly tombstones in the graveyard—a plan slab, sometimes of native sandstone or slate, a narrow one if placed upright, and wide if placed horizontally, and sometimes a small monument of marble, with an inscription, frequently accompanied by a line of poetry, marked the resting-place of the dead. These country graveyards grew up then, as they do now, with brush and briers, but it was customary for the neighbors to meet once every two years and clean them up, making needed repairs and filling up the graves that had settled in the meantime.

Birthdays were not much regarded in my young days, except by the children. They, like children now, were anxious to get old fast, looking forward with glad anticipation to the time when the girls would be eighteen and the boys twenty-one, when they were said to be "of age." We had fewer holidays then than now. We celebrated the Fourth of July, never working on that day unless there was some danger of losing the wheat harvest. It was celebrated more in the way of picnics and family gatherings than by Fourth of July orations. In fact, I can not remember that I ever heard a Fourth of July oration until after I was twenty-one. Some small respect was paid to Washington's birthday in the larger towns, but not in the country. Thanksgiving was usually celebrated by a good dinner at home, sometimes a turkey, sometimes not—but with nothing like the elaborateness of the New England Thanksgiving. There was religious service in the towns. A few years before the breaking out of the Civil War, the preachers availed themselves of this opportunity to express their views on national subjects. The people of the opposite party regarded this as "preaching politics," and did not feel under any particular obligation to attend. In the country we usually went hunting or attended a turkey-shoot or raffle.

Not very much attention was paid to Christmas. We Calvinists had an idea that an elaborate celebration of Christmas indicated a leaning toward popery. We usually had a good Christmas dinner, however, and if there was a turkey-shoot in the neighborhood, it was all right to go and take a shot. We paid little or no attention to New Year's Day except to settle up book accounts with the store-keepers or the neighbors. The ordinary farmer who went to the store to settle up his account and found there was nothing coming to him

on account of the butter and eggs that his farm had furnished during
the year, was rather out of sorts, but felt happy if there was a comfortable balance due him for the produce.

The presents given him to the children at Christmas would sur-
prise you by their meagerness and simplicity. There were no wax or
bisque dolls, no dolls that could talk or cry, or that needed elaborate
clothes. If a little girl wanted a doll, she generally made it herself,
but she thought quite as much of it as of an expensive one. In fact,
I think she thought a good deal more of it, even though it was made
of a Kershaw squash or rags. In either case, it served as an expression of the latent instinct of motherhood natural to the female, without regard to race, wealth, education, position, or anything else.

The birth of a child in the family was regarded as a great event,
especially by the older children, though there was not very much
said about it. I remember when my sister Margaret was born. It
was in haymaking time. My father had a very intimate friend who
neighbored with him a good deal, and I wondered the next morning
that he did not say anything about the new baby. We were raking
hay, I leading and the two older men following. My father said
nothing about it till about half-past ten o'clock, and then said:

"George, we had an addition to the family last night—a girl."

George answered: "The usual good luck, I suppose?"

My father said, "Yes," and that was all there was of it. It was
assumed that children would come to every family about every
other year, or at least every third year, that everything would go
well, and that mother would be up and about in nine days, or in ten
at the farthest. Of course, the neighbor women came to look the
new arrival over. The old grandmothers put on their spectacles and
said: "Lawsy me! Which side of the family does she take after?" on
which point there was very naturally difference of opinion. For my
part, I never could see that the baby took after either side.[144]

My Dear Great-Grandchildren:

The small college, whether in the nineteenth or the twentieth
century, has this great advantage over the large one: that the pupils
can get in closer touch with the professors. For obvious reasons,
the small denominational college can get a bigger man for the same
salary. If the small college has even one big man, the student is
likely to get more benefit from this than from an even bigger man in

a larger college, for the simple reason that it is easier to get in touch with the best thing in the college, a Man. In the larger school, almost the entire instruction must be given by men of little reputation, men of whom the student has probably never heard until he entered the classroom. He seldom gets very close to the man of reputation connected with the school.

I was very fortunate. When I went to Geneva Hall, I was thrown in close touch with some really big men. The biggest of them was a man long since dead named Dr. Sloan, then about forty years of age. He was bred in the east, had evidently received a very thorough training, and had moved in the very best society. He was a thorough teacher, and took a lively personal interest in every boy connected with the school. He knew how to get work out of boys without formally requiring it of them. I remember very well the first day I met him. They were organizing the classes at the beginning of the session. The class of which my recollection is most vivid was that in Latin, textbook, Caesar. I knew a little Latin grammar, but did not know that thoroughly, and one of the professors said that I could not keep up with the work of that class. Fortunately, there seemed no place else to put me, and Dr. Sloan said: "Put Wallace in; he'll get through all right." I felt greatly encouraged by that. The first day the doctor sat and looked at us a bit, and, as I was at the end of the seat, he said:

"Wallace, what did you come here for?"

I stammered, after a time, "To get an education."

"An education—what is an education?"

After he had, so to speak, chased me around the room with that sort of a prod, he said: "Well, what do you want with an education? Then, after a while: "How do you expect to get it?" He wound up his examination of each one of us—for he treated us all alike in order—with: "What do you want to do with an education when you get it?" And that was the first day's recitation in Caesar!

I wondered what would come next. The next day, he simply asked one of the boys to hand him a book, and motioned for us to begin translating. If there was a wrong accent, he twisted a lock of hair which hung behind his right ear; if a wrong quantity, he frowned: if a wrong pronunciation, he gave a peculiar stamp with his right foot. We came in a little while to the third sentence in the first book of Caesar. It went around the class, and nobody could translate it. He simply handed back the book and said, "That will do until tomorrow," and when we came back the next day he began with that sentence. That was his method of teaching Caesar. I do not know how the other boys felt about it, but I made up my mind that I would have that lesson if I sat up all night to get it, and I had it. Only in extremely difficult passages did he give us any direct help. We had to get it ourselves.

Then he had us visit him in the evening, one or two at a time, and talked with us about our parents and our aims in life. I was a very awkward boy in those days, and not very particular about my clothes. His wife, a very refined lady, and evidently accustomed to the best society, one evening called my attention to this and said I would succeed a great deal better if I would be more particular about my appearance. He said, "Well, that's all right. She is giving you good advice, but I would advise you not to neglect your studies to take in polish. If you master your studies, there will be something to polish, and that will come by and by."

Then we had Doctor Milligan, who I afterwards learned during a long friendship lasting until his death, was only four or five years older than myself. He was a most genial man who knew how to enlist the boys' sympathies, and was constantly pointing them to higher things. Other professors we had—good ones, very good drillmasters who insisted on thoroughness—but they failed to grip me as did these two men.

The course of study was not an extensive one, nor nearly so large nor so long as in the best high schools of the early part of the twentieth century, and was mainly adapted to fitting men for a professional life. Outside of the professions there was nothing vocational about it. English grammar, composition, higher arithmetic, algebra, plane and spherical geometry, trigonometry, Latin, Greek and Hebrew—the latter in the senior year. These and work in the literary societies were the main features of the education we received in that school. The main object was to make preachers, and an unusually large number of the students became preachers, and are sporting "D.D's" and "L.L.D's" to this day.

Education was cheap in those days, and perhaps it may be regarded from the modern viewpoint as rather cheap education in itself. I do not think I ever paid as much as four dollars a week for board and room—usually two and a half or three. The board was usually good, and I wish I could enjoy now as hearty a breakfast as I did at my first boarding-place, where they had splendid flapjacks, pancakes swimming in genuine maple syrup made on the farm. We had these nearly every morning, or else buckwheat cakes (home-grown) hot from the griddle, and plenty of good butter, country-cured ham, and fresh eggs in season.

The rooms were bare, seldom carpeted. A washbowl and a pitcher served for all ablutions, and we invariably threw the dirty water out of the window, for there was no sewer, and this was the easiest way of disposing of it. I shall never forget one incident while I was at Jefferson College, two years afterwards, and rooming in the second story. I used rainwater which had a plentiful supply of soot in it, being in coal county, and when I threw the dirty water out of the window, as usual, it came squarely on top of the bald head of my landlord, who happened to be passing under that window at the time. I saw him after I tipped the basin, and began profuse apologies. He must have seen a saint of a ripe degree of grace, for he only looked up and said; "Did I ever!" I do not think, however, that he ever got rid of a suspicion that I did it on purpose—which I did not.

The streets of the little village of Northwood were simply country roads poorly worked; the sidewalks simply one oak board, about a foot wide, on which men usually walked alone, but on which I observed that a boy and a girl could walk together and balance well.

I remember but one case of disciplining, involving, however, two persons. There are cranks in colleges as well as elsewhere, and we had one, a queer fellow, who wore in summer a hat of straw with as wide a brim as the broadest of the ladies' hats in the year 1910. This brim was held up by a string on each side. In the summer he went barefooted and wore a gown of some sort which reached down to his feet. He argued with the boys that it was wrong to take the life of any animal, even that of the most offensive insect. He argued also that we should eat nothing that not was grown in our own immediate environment. Hence, he abjured tea, coffee, spices and sugar, except maple. He refused to cut either his hair or his beard. He was fitting himself as a missionary to Afghanistan, and we all wished he were there. He made some slighting remarks about the girls with whom two of the older boys were going, and naturally incurred their special enmity. On election night, the fourth of November, 1856, these boys waylaid him as he was passing through a vacant room in the college, which had not been swept that summer. There was a struggle, and much hair and gore were found on the floor in the morning. In taking off his long hair, which he had refused to even have cut, they also took some scalp with it. Of course, an offense of this kind could not be overlooked, and the boys were suspended, but they had no difficulty in entering another college of the same grade, or a higher one, in the same classes.

Longfellow had made Hiawatha famous at that time, and many

were the sheets of paper spoiled by descriptions of this tragic scene in Hiawatha style. I spoiled a number myself, but can only remember the beginning of one: "On the fourth night of November," in which were references to the silvery moon, a sky overcast with clouds, and untrodden snow, and stars as they looked down upon the scene.

There were few amusements as an outlet for the pent-up energies of the students; no baseball, no football, no tennis, no gymnasium. There was an occasional visit with some one of the boys at his home in the country, a visit to the nearest town, and a supper at the hotel with some girlfriend, a day's hunting for wild turkeys, which were not numerous, and for squirrels, both the black and the gray—these furnished about all the diversion we had.

Generally speaking, we were hard-working, earnest students. The second year, I was in the habit of getting up at four o'clock, which I would not like to do now. I studied till six then had breakfast and an hour's walk if the roads were dry or frozen. There were recitations in the forenoon, an hour for dinner, some recitations in the afternoon, an hour for supper, and then study till ten. We have made much of our literary societies, and fierce was the rivalry between students for society as well as class honors. I was once foolish enough to rejoice over the defeat of one of my rivals by declaiming in my turn at "Woolsey's Lament," with variations: "Adieu, a last adieu to all my greatness!" At the next night he, with similar variations, declaimed something about Napoleon's decline, fall, and banishment to Elba. It taught me a lesson, "not to rejoice when thy enemy falleth." I richly deserved the unmerciful scoring I received.

I, with some of the other boys, had been on a sleigh ride to a little town called Roundhead. On the way home, our sleigh broke down, and we borrowed a log chain from a farmer without asking his consent, intending, of course, to return it. The farmer was furious, and came to town the next day, demanding our arrest. Of course we settled with the farmer, but my friend, in his declamation," substituted "his fight from Roundhead"! I happened to censor that evening, and managed to pay the highest compliments to my castigator, much to the amusement of the boys.

And yet, looking back over more than a half century, I feel that I owe this fellow a good deal. He and I never did get along. One day he said something particularly mean about me at the boardinghouse, which rankled even when the family were at worship; and then and there it occurred to me that I would take advantage

of the fact that a room had been left vacant by one of the students who had been suspended, and change my boarding house, with this further advantage that I would have a room to myself. While there I became acquainted with a young lady much older than myself, who was boarding at the same house, and through her, two years later, I had the opportunity of becoming acquainted with the lady who became my wife. Thus do the most trifling things, to which we pay no attention at the time, shape our destiny. Some call it chance, some luck; I call it providence. Just think of what a difference it might have made in the Wallace family if I had not fallen out with this schoolmate over a trifling remark, changed my boarding-place, or perchance married a less noble woman than your great-grand-mother!

Last spring, the *American Magazine* saw fit to publish a sketch of my life. A few weeks afterward, I received a letter from a professor in a college out in Utah, stating that he supposed I was the same chap who took part in a chicken dinner in the room of one of the students, the chickens for which had been "borrowed" (like the log chain) from a farmer whom the boys greatly disliked. He called my attention to the moot court which was held on the college campus on a dark night, in which he says I acted as a judge, arraigned the culprits, sentenced them to deliver to the farmer resolutions of regret that anyone connected with the college had been guilty of interfering with his feathered flock, promising that it should never occur again, if the discipline of the students could prevent it, and that they should pay him one dollar—more than twice what two tough old hens were worth then. I had really forgotten all about that moot court, but have an indistinct recollection of the feast, the chickens for which I suspect were feloniously purloined! I presume that human nature in college is now, and always will be, much as it was then and always has been.

My Dear Great-Grandchildren:

You may have noticed already—and if not, you will notice it before you get very far along in life—that often the little things, things apparently trivial, sometimes mere accidents, lead to very large results, and in fact often change the whole course of life and destiny—just as a switch on a railroad track may shunt the traveler off into a track going into a different and by and by opposite direction—sometimes to his death.

I have noticed this all through life. In fact, my career has more than once been changed by things that seemed trivial, as, for example, my first meeting with your great-grandmother. You want to know the details of that, but I won't tell you.

In the year 1902, I was delivering some lectures at a farmers' institute at Winchester, Illinois. Some progressive farmer had furnished the boys in that county with samples of improved seed corn, and had offered a prize for the best corn, and samples were being shown. The institute management had also offered prizes for the best samples of corn grown by the farmers generally. The superiority of the samples shown by the boys over those of the fathers was to me somewhat surprising. In fact, the older men "were not in it."

The samples of corn were sold, apparently to pay the expenses of the institute, and the farmers were trying to buy the boys' samples at perhaps twice the price of corn on the market. I finally rose and told them I would not stand for that, and that if they did not pay the boys a decent price for their corn, I would buy the whole lot and send it to Iowa, where I knew there were men who would appreciate that kind of corn. This forced them to pay the boys a fair price. I really did not want to take the corn to Iowa, because it was too large a variety for our latitude.

On the way home, it occurred to me that *Wallaces' Farmer* might run a corn contest offering prizes. My sons took favorably to the idea, and we divided the country into three sections—south, middle, and north. The southern division, as I recall it, was that section south of the line of the Chicago, Burlington, and Quincy railway in Iowa; the northern, north of the line of the Illinois Central, extending from the Atlantic to the Pacific. The middle division was between the lines of these two railroads.

We decided that we would furnish a pint of the very best seed corn of the variety adapted to the division in which he lived to any farm boy under nineteen who would hustle around and send us new subscribers in clubs of three to five, and that we would give to the winners in each division prizes ranging from one hundred dollars down. We had clubs sent in from as far as New Hampshire on the east and Colorado on the west, as far south as Arkansas, and as far north as Minnesota and southern Wisconsin.

In the southern division we gave them their choice of Boone County White (a large white variety) and Nlm's Legal Tender, which was a very popular variety of yellow corn at that time requiring about the same number of heat units for maturing as the Boone

County White. For the middle division we offered Reid's Yellow Dent and Leaming, both requiring about the same number of heat units. There was not any improved corn at that time in the northern part of our territory except Minnesota No. 13, a yellow corn and at that time not very greatly improved. We used this and Longfellow's Yellow Dent and Pride of the North. The experiment stations were merely making a start in getting varieties of improved corn.

We had just put up our new building at the corner of Eleventh and Walnut streets. We had an upstairs room about 44 x 70 feet, and placed the exhibits of corn sent in according to latitude, the south end of the room being given over to corn from the south; the middle to that from the middle division, and the north to that from the northern section.

It made a very pretty corn show, and in walking thru the room from north to south, one could get a good estimate of the capacity of the different sections of the country for growing corn, and the types of corn they could grow best. A good many visitors came to the building. It was really the first attempt in a large way to improve corn in the Mississippi Valley by giving an exhibit, and particularly of the different types of corn adapted to the different sections.

Our boys' corn contest did an immense amount of good. It resulted in introducing better varieties of corn, especially in the northern district. But, best of all, it got a lot of boys really interested in the farm. Many of these boys later went to the agricultural colleges, and have become men of note as farmers, breeders, agricultural teachers, or newspaper men. Some of the letters I prize most highly are from boys who have written to thank us for these corn contests, saying that they marked the beginning of real life for them.[145]

Part Four

Letters from *Tributes to Henry Wallace*

The last photo of Henry Wallace, taken less than a month before his death. It shows "Uncle Henry" holding his first great-grandchild, Henry Browne Wallace, then about four and one-half months of age. In the center is his son, Henry C. Wallace, and at the left is his grandson, Henry A. Wallace. Four generations of oldest sons.

A Letter from the Sons of Henry Wallace, Wallace's Farmer March 3, 1916, to the readers of *Wallaces' Farmer*

Swift as a flash of His own lightning, the sword of the Almighty at one stroke has severed the threefold tie which bound father and sons, business associates, and intimate friends and companions.[146] We have lost all in one.

In such an hour, the written word is a weak vehicle for the emotions that sweep over the soul and flood it in a sea of inexpressible grief. But the consciousness that thousands of sincere friends are grieving with us, and would willingly share our burden, strengthens us to speak as best we may at this time of the father who no longer is with us.

As a father, he was all that a father should be: loving, kind, strong, tender, thoughtful, just. He came of a breed which believes in training up children in the fear and admonition of the Lord—but his way was by instruction which children could grasp, by precept and example, not by command and stern rules and laws. Harshness was utterly foreign to his nature. So far as we can remember, he resorted to corporal punishment but once, and the occasion became a historic event in the family. He could think the thoughts of a child; he could see with the eyes of a child; he had the sympathies of a child. So with him, children, both his own and others, always were unafraid.

As his children grew, he grew with them. He shared their sorrows and their joys. Throughout his life, he had the rare faculty of meeting each child and grandchild upon his own intellectual plane, of finding pleasure in its pleasures. And so he guided while walking beside us. He won both the respect and the confidence of all his children. As we came to the age of discretion, he recognized us as independent personalities, as men and women who should be capable of shaping their own affairs, and who must accept their own responsibilities. By insensible degrees, he became elder brother and friend as well as father. If something done or undone met with his disapproval, never was there harsh reproof. We became conscious of his disapproval sometimes in one way and sometimes in another, depending upon the needs of the occasion. But never, in all our lives, was his correction or reproof administered in a way which created a feeling of resentment or the slightest break in the dual relation of father and friend. To him, a quarrel between father and son was a terrible thing, and his most widely read book, *Uncle Henry's Letters to the Farm Boy,* was inspired by just such a quarrel which he chanced to witness in a friend's home.

It is not unusual for sons and fathers to be associates in business. It is one of the common relations of life. As sons come to years of maturity, they are trained into the business of their fathers, and succeed them in the natural course of events. But ours was an unusual relationship. When the time of trouble came to him, twenty-one years ago, we had not been working together. The father had been engaged in the publishing business with others. One son was in college work, a second had just reached man's estate, and a third was still a boy in high school. *Wallaces' Farmer* was started at that time. The father and the two older sons came together in the effort to repair the family fortunes, to vindicate the family name, and to establish a farm journal which could always be depended upon to stand for certain principles, which soon afterward were expressed in the words that have ever since been carried on our front page, "Good Farming, Clear Thinking, Right Living." It was, therefore, not a case of the sons following into the father's business, but of father and sons coming together to establish a business of their own. And with all of us, the motive was higher than that of mere money-making. In fact, from a business standpoint, the enterprise seemed very difficult.

The father had had long experience in newspaper work, altogether on the editorial side. The sons undertook the business side, in which neither had had experience worth speaking about, although the older had spent his spare time as a boy and his vacation periods in the mechanical department of a printing office. Under such circumstances, nothing would have been more natural than that the father should expect to direct the enterprise. He did nothing of the sort. From the beginning, he expected that success would come, and he believed it would come through each man doing the work he could do best. He devoted himself exclusively to the editorial work. He knew that we lacked experience. He knew that we were young, and would make mistakes. But he made us feel that we had his entire confidence. Always ready with counsel and suggestion when asked, he left us entirely free to do our work as we thought it should be done. He expected us to grow as the business grew, to profit by our own mistakes. Never in the twenty-one years was there from him censure or fault-finding for things done or undone—nothing but encouragement, confidence, good cheer. None but a really big man can do as he did under such conditions.

And as he was as a father and as a business associate, so he was as a friend and a companion. He had a most comprehensive fund of information, covering almost every subject of human interest. A long railroad journey was never tiresome with him; soon he would find the most interesting man on the train, and time would pass swiftly in gaining and imparting ideas and knowledge. Never was there a more genial companion; rarely a more charming conversationalist. It was not necessary that you should talk about what he wanted to talk about. He would talk on the subjects in which you were interested. All he asked was that the conversation should be about something worthwhile. Whether his companion was a president, a farmer, a banker, a minister, a merchant, a manufacturer, or a day laborer, he always found the plane upon which they could converse as equals and with mutual benefit. His ability to become interested in the things which interested the various people he met had much to do with keeping him young. On the day he died, his mind was as vigorous, and his interest as keen in current events as it was at any time in his entire life.

He had a most delightful sense of humor. He was full of anecdotes, and no one found more enjoyment in a good story that had a point to it. Unlike many, he could see the funny side of a joke on himself, and in genial give-and-take chaffing he was a master hand. He was incapable of taking offense when none was intended. He took it for granted that you were his friend, and until he found himself to be wrong in this, no explanations were necessary to account for what, to the supersensitive, might seem to be a slight.

Some of our most treasured memories cluster around our companionship with him, not as a father or as a business associate, but as faithful friend and companion, as host or guest in the home, or visitor in the office.

In all these relations, his understanding of human nature was profound. His sympathy was all-embracing. His charity was unbounded. He had touched so many sides of life in his career as minister, farmer, editor, lecturer, that he intuitively sensed the point of view of the persons with whom he was talking. He could share their joys and sorrows. When some of the young folks of the *Farmer* office decided to set up homes of their own, he performed the marriage ceremony. When death entered their homes, he conducted the funeral services. One of the keen enjoyments of his last week was the success of his strategy in bringing to a happy conclusion the love affair of a girl friend. It was a delight to him to help in time of trouble. His office long since became a place of refuge to those

who sought help and sympathy. And he was the safe repository for the most intimate correspondence from many who never met him face to face, but, from having read his writings, knew he could be trusted.

Seldom did he offer advice on the personal affairs of life, but gave it freely when asked. With gentle tact he drew out the whole story, and then, putting himself in your place, he went over it, examined it from all sides, and suggested possible ways out of the difficulty. If you were wrong, he did not tell you so; gently he led you to the point where you could see it for yourself.

His charity for human frailty was unbounded. The stanzas from Burns which he most often quoted were:

> Then gently scan your brother man,
> Still gentler sister woman;
> Tho they may gang a kennin wrang,
> To step aside is human.
> One point must still be greatly dark,
> The moving why to do it;
> And just as lamely can ye mark
> How far perhaps they rue it.

This charity grew upon him as his life mellowed and ripened. Never could he condone wrongdoing. But he could not sit as judge. His thought always was to bring back the wrongdoer to ways of right-doing, and in such a way that it would be harder to go wrong a second time. He felt that those could best serve their fellow men who recognize human weakness, and the temptations which beset it. He had no hope that the world could be reformed in a day. His preaching was "line upon line, precept upon precept." His way to make the world better was to make known the rewards of well-doing as well as the penalties of evil-doing; to give to the life a purpose which would make it easier to overcome temptations which can not be removed. And, above all, to teach the responsibility of man to his maker. For he had learned from long personal experience that the source of all the power worth having comes from an abiding belief in and practice of the teachings of Jesus.

He never demanded either love or respect. He won both from everyone with whom he came in contact. He made friends because he was friendly. The vigor with which he stood for fairness and justice to the farmer made him many adversaries. As with all men of his type, he was subjected to the narrow criticism of men who lacked his vision, who could not grasp the unselfishness of his purpose. This never worried him. In personal contact he disarmed them with a gentle courtesy that was simply irresistible. Tenacious in his adherence to great principles of right and justice, he could advocate them with the greatest tolerance for the opinions of those who might differ from him. Many of his warmest friends and admirers in his later years were men who had been his most severe and intolerant critics. And the letters of sympathy and understanding which are now coming to us from these men are greatly treasured. He could on occasion become filled with righteous indignation, but he was incapable of resentment, or brooding over wrongs unjustly inflicted.

What we have said is not intended as an estimate of the man nor of his works. In the not distant future this will be undertaken in an adequate manner. We have tried simply to give utterance to the many thoughts which surge upward, seeking expression. And if our words seem feeble and ill-chosen, we ask our friends to remember that they are written at a time when we are bowed down under the greatest burden of grief which has ever come upon us.

What of the future?

Anticipating this very question, he himself answered it in *Wallaces' Farmer*, in the issue of February 18th, the week before his death. Under the heading, "*Wallaces' Farmer* Comes of Age," he wrote of the work of the past twenty-one years. In the last paragraph of this article, he said:

"When the present editor passes on, it will make no change in the policies or principles for which *Wallaces' Farmer* stands. We ought to be, like the young man coming of age, only in the beginning of our usefulness and power."

So far as in us lies, we will justify this last public expression of the confidence our father reposed in us. We take these words for our own. As our father stood, so shall we.

We shall miss his ripe counsel. We shall lack his mature wisdom, his facility of expression. But our willingness to serve, our integrity of purpose, our fidelity to the farmers of the West, shall be as firm as his own. And the privilege of having had him for so long as guide, philosopher, and friend will help us in this. Our ambition shall be worthily to inherit that trust and confidence so freely given to him.

For many years, the burden of editorial work has very largely been shared by those who now succeed him. In these later years, as his interest grew in the larger movements for agricultural betterment, he has been free to come and go just as he wished. He ceased to hold himself responsible for the week-by-week task of filling the paper. He wrote what he felt like writing, and no more. There were editorials of a certain kind which no man but Henry Wallace could write. After a time, these will be missed. But the responsibility for most of the editorial work, he long since shifted to younger shoulders. Relieved from routine work, he took up with great zest some special articles which are safely stored away, and these will appear in due season.

In the minds of thousands of our readers, there will be concern about the weekly Sabbath school lessons, which for many years have been perhaps the most popular single feature of *Wallaces' Farmer*. These fears may be set at rest. These lessons will be continued, and they will still be "Uncle Henry's" lessons. During the years that have gone by since this work has begun, the scriptures covered by these international lessons have been gone over three times. Occasionally, a new lesson has been introduced, but very rarely. So from week to week, in the time to come, "Uncle Henry," in his own words, will continue to interpret the Holy Writ, and apply it so wisely and so helpfully to the needs of the average man, woman, and child. No other portion of his work will be more enduring, nor remain perpetually so fresh.

And now, deeply conscious of the rich heritage which has come into our keeping, fully sensible of the responsibilities which we must assume, cheered by the words of sympathy which come in every mail, we set our faces to the future.

Of our fidelity to his standards and ideals, we invite the most unsparing criticism. Of the ability with which we maintain them, we plead for the kindly charity which he gave in such generous measure.

He looked upon *Wallaces' Farmer* as his monument. To keep it clean and pure and worthy is the task which we assume. To make it an ever-living, growing testimonial to his life and ideals is a work in which we must have the help of his friends and ours.

And, as in the past so in the future, the purpose of *Wallaces' Farmer* shall be the advocacy of Good Farming, Clear Thinking, Right Living—no one alone, but all.

His Sons

A Letter from the Daughters of Henry Wallace, from *Wallaces' Farmer*, Hearts and Homes Department, March 10, 1916

I can not say, and I will not say
That he is dead. He is just away!
With a cheery smile and a wave of the hand,
He has wandered into an unknown land,
And left us dreaming how very fair
It needs must be, since he lingers there.

On the evening of February 22nd, it seemed for a time as if the world had rushed on, leaving us utterly alone. One minute, the daughter whose very important share in the success of *Wallaces' Farmer* was the making of the home for its editor, was wondering how father had stood the day. Teddy, the canary, swayed back and forth in his swing, tilting forward to look inquisitively at the unpressed pillows of the empty couch; the reading lamp lighted by the stand, holding the well-thumbed Bible; the dressing jacket in readiness in the hall, and the easy slippers side by side on the floor; the house tuned for his coming.

The minute that followed brought the crushing news that the father we loved had passed from the earthly home forever. What had been eloquent witnesses of his expected presence, became heartbreaking reminders that he was gone.

We know he would have chosen to pass out of our lives in the beautiful way in which God took him. We know he would be grieved if we let his absence hinder us in the day's duties. Yet you who have known him, who have sought his written and spoken word, will understand what his departure must mean to his children. Father went with his father, mother, brothers, sisters, children, and wife to the grave, but he did not let them out of his life. We have taken him to the grave, but his beloved influence will remain with us, and his ideals shall guide us. The words of sympathy which are coming to us from so many dear friends have greatly comforted us.

His Daughters

Compositions by Wallace School Pupils from *Tributes to Henry Wallace*

By Gladys Cram, Class A, Eight Grade

Temper is sometimes excusable in a boy or girl, but never in a man or woman.[147] We should begin now to control our temper. One morning at the breakfast table, Henry Wallace did not like something, and on being reproved, lost his temper. His father threw water in his face again and again, and by the time his father had ceased, he had gained control of himself, and there and then he decided that he could and would control his temper. He told a story of a man who whistled every time he felt his temper rising, and in that way learned to keep calm. "Count ten before you speak" is rule commonly given to help control your temper, but very few observe it. Henry Wallace says: "If you can be rightfully angry about a wrong or an injustice, and control yourself, it marks the really strong man or woman."

By Hugh Gallagher, Class A, Eighth Grade

"The boy who is not a reader in this day and age of the world, is very likely to be a nobody," was predicted by Henry Wallace of the coming men. "Uncle Henry" recommends that a boy first read the Bible. If he desire to speak in clear, simple, forcible English, he will find a helper in Bunyan's *Pilgrim's Progress*. A lad ought to thoroughly master his school books, and read and reread a few excellent books, not only for pleasure, but for obtaining knowledge. A good newspaper that clearly sees both sides of a question, and a condensed current magazine, will also greatly influence his later life.

By Edna Chine, Class A, Eighth Grade

A well-developed boy should have first of all a dog—a bright, honest and industrious dog, that can look you squarely in the eye yet without flinching, and fight for you when it is necessary—and then a chum. A good chum should have good qualities. He or she should be loyal at all times, clean-minded, industrious, and economical. A boy or girl who can not control his or her temper is not the right type of a playmate. He should have good blood and manners, and not be profane, quarrelsome, or easy to "fly off the handle," as is sometimes said. A good rule for a boy to follow is expressed in a quotation of "Uncle Henry's": "Choose as your

chum a boy that respects his father, loves his sister, fights for his little brother, and adores his mother."

By June Gray, First Grade

My name is June Gray. I go to Henry Wallace School. I am in the first grade. Our school is named for Mr. Henry Wallace. He was a good man. We loved him. He loved good boys and girls. We want to be the kind of boys and girls he loved. Monday is his birthday. We are going to have a party on his birthday.

From Henry Wallace's Last Will as excerpted in *Tributes to Henry Wallace*

There are some possible temptations against which I must warn you.[148] The family has been prospered in many ways beyond anything which I hoped or could have expected. It has prospered in a material way, and enjoys a reasonable measure of public confidence because we have never sought wealth nor office, nor social position, as ends in themselves, but merely as a means of enlarging our possible usefulness to the community at large. Any serious departure from this policy will be fatal to the best interests of the family. I see no indication as yet. But the temptation to amass wealth, the temptation to gain position, political or social, for purely selfish purposes, come naturally with prosperity. Avoid all this. Keep clean in speech, clear in mind, vigorous in body, and God will bless you.

I am aware that it is quite unusual to discuss such matters as I have dealt with in this, and in the preceding sections, in a last will. I have departed from the usual custom for the reason that I regard any help that I may give my children and grandchildren in the supreme work of life, that of living worthily, is of far greater real value to them than any worldly possessions I may hand down to them."

Having acquired a competency several years ago, I have, as my children all know, contributed to some worthy object and divided the rest among them each year, thus distributing my annual income, save that needed for a life of comfort. I have found great satisfaction in this, and advise them, after seeing their children well started in life, to follow my example. It will save them from many temptations, and give them a saner view of the real object of life.

HENRY WALLACE
BORN MARCH 19, 1836
DIED FEBRUARY 22, 1916

About the Editor

Fourth-generation Iowa farmer's son Zachary Michael Jack has authored or edited many books, including several award-winning volumes examining the relationship between agriculture, environment, and education, most recently *Love of the Land: Essential Farm and Conservation Readings from an American Golden Age*. His edited collections *Black Earth and Ivory Tower: New American Essays from Farm and Classroom* and *The Furrow and Us: Essays on Soil and Sentiment* have both been nominated for the Theodore Saloutos Award for the year's best book on agricultural history and have been featured in the *Chicago Tribune*, the *Des Moines Register*, and on Chicago and Iowa Public Radio.

Zachary's interests in environment and community originate in his family's more than 150-year-old Iowa Heritage Farm and timber, and find expression in his directing and founding of the Iowa School of Lost Arts for children in 2004. Jack's Pushcart Prize-nominated collection *Letters to a Young Iowan: Good Sense from the Good Folks of Iowa for Young People Everywhere* continues a commitment to midwestern youth begun in his days as a children's librarian in Ames, Iowa. An advocate of writing "in place," Jack is an advisory board member for the Interversity Place Studies listserv, a former section editor for an Iowa community newspaper, and a former writer-in-residence at New York's Blue Mountain Center, Ireland's Tyrone Guthrie Centre, and Mexico's Great River Arts Institute. An assistant professor of English, Jack teaches courses in writing and rural studies and urban studies at North Central College.

Notes

1. Wallace, Henry, *Letters to the Farm Folk* (Des Moines: The Wallace Publishing Company, 1915), 46-47.

2. Ingham, Harvey, "Henry Wallace as Editor and Publisher." In *Tributes to Henry Wallace* (Des Moines: The Wallace Publishing Company, 1919), 5.

3. Ibid.

4. *Des Moines Register and Leader*, "Eulogies for Henry Wallace at Y.M.C.A. Meeting," 1916. In *Tributes*, 22.

5. Beckman, F.W., "Henry Wallace's Work for Farming and Farm Folks," *Iowa Agriculturalist*. In *Tributes*, 48.

6. Wilson, James. "The Influence of Henry Wallace on Farm Life." In *Tributes,* 6.

7. Wallace, Henry, *Letters to the Farm Folk*, 18.

8. Ibid., 20.

9. *Proceedings of the Second National Conservation Congress at St. Paul*, September 5-8, 1910 (Washington: National Conservation Congress, 1910), pp. 188-94.

10. Ibid.

11. Ray Stannard Baker, "Interesting People," *The American Magazine*, 71 (December 1910), 175.

12. Ibid., 176.

13. *Proceedings*, 303-06.

14. Quick, Herbert, "Good Farming, Clear Thinking, Right Living." In *Tributes*, 73.

15. Kirkendall, Richard S., ed., *A Documentary Profile of the First Henry Wallace* (Ames: Iowa State University Press, 1993), 211.

16. Ibid., 220.

17. Ibid., 219.

18. Ibid., 154.

19. Hsu, Caroline, "The Greening of Aging," *US New and World Report*, June 19, 2006.

20. Public Broadcasting Corporation, "Texas Ranch House," http:// www. pbs.org/wnet/ranchhouse/about.html (accessed December, 2006).

21. "How Heading to the Woods Can Heal a Child," *Toronto Star*, A05, June 11, 2006.

22. Ibid.

23. "A Great Farm Editor," *The American Review of Reviews*, 53, (May 1916): 607-08.

24. Quick, 75.

25. Lord, Russell, *The Wallaces of Iowa* (Boston: The Riverside Press, 1947), 7.

26. Ibid.

27. Johnston, W.P. "The Influence of an Earnest Teacher On the Life of Henry Wallace." In *Tributes*, 63.

28. *Des Moines Evening Tribune*, "Hundreds at Bier of 'Uncle Henry' Wallace," February 1916. In *Tributes*, 13.

29. *Des Moines Register and Leader*, "City Pays Tribute to Great Citizen," February 1916. In *Tributes*, 14.

30. Wallace, Henry, *Uncle Henry's Own Story of His Life*, vol. 2 (Des Moines: The Wallace Publishing Company, 1917), 27.

31. Hurt, R. Douglas, "Series Editor's Introduction." In *A Documentary Profile of the First Henry Wallace* (Ames: Iowa State University Press, 1993), xi.

32. Wallace, Henry, *Uncle Henry's Own Story*, vol. 1, 92.

33. Ibid.

34. Ibid., 109.

35. Ibid., 106.

36. Lord, 48.

37. Wallace, Henry, *Uncle Henry's Own Story*, vol. 2, 27.

38. Lord, 56.

39. Ibid.

40. Ibid., 57.

41. Wallace, Henry, *Uncle Henry's Own Story*, vol. 2, 81.

42. Lord, 71.

43. Wallace, Henry, *Uncle Henry's Own Story,* vol. 2, 98.

44. Lord, 72.

45. Wallace, Henry, *Uncle Henry's Own Story*, vol. 2, 110.

46. Ibid.

47. Lord, 78.

48. Ibid., 79.

49. Ibid.

50. Henry Wallace, *Clover Culture* (Des Moines, The Homestead Company, 1892), 5.

51. Lord, 81.

52. Ibid., 83.

53. Ibid.

54. Ibid., 84.

55. Ibid., 90.

56. Ibid., 93.

57. Ibid., 99.

58. Ibid., 103.

59. Ibid., 100.

60. Wallace, Henry, *Uncle Henry's Own Story*, vol. 1, 48.

61. Lord, 113.

62. Ibid.

63. Wallace, Henry, *Uncle Henry's Own Story*, vol. 3, 64.

64. Lord, 120.

65. Ibid, 126.

66. Wallace, Henry, *Uncle Henry's Letters to the Farm Boy* (Des Moines: The Wallace Publishing Company, 1897), 142.

67. Lord, 126.

68. Ibid,, 128.

69. Ibid., 129.

70. The Wallace Publishing Company. *Tributes to Henry Wallace.* Des Moines: The Wallace Publishing Company 1919, 34.

71. Lord, 131.

72. Ibid.

73. The Wallace Publishing Company. *Tributes*, 35.

74. Lord, 130.

75. Ibid., 131.

76. Wallace, Henry, *Uncle Henry's Own Story*, vol. 3, 77.

77. Ibid., 74.

78. Ibid., 71.

79. Ibid., 74.

80. The Wallace Publishing Company. *Tributes*, 35.

81. Lord, 133.

82. "Receipt of No. 2 of Vol. 1 of *Wallace's Farm Library*, 'Trusts and How to Deal With Them' by Henry Wallace," *Milwaukee Journal*, pg. 6, col. B, October 21, 1899.

83. Lord., 138-39.

84. Ibid., 141.

85. Wallace, Henry, *Uncle Henry's Letters to the Farm Boy*, 90-91.

86. Wallace, Henry, *Letters to the Farm Folk* (Des Moines: The Wallace Publishing Company, 1915), 29-30.

87. Lord, 143.

88. Ibid., 145.

89. Bowers, William L., *The Country Life Movement in America, 1900-1920* (Port Washington, NY: Kennikat Press, 1974), 24.

90. Ibid., 25.

91. Ibid., 26.

92. Lord, 151.

93. Ibid.

94. Ibid.

95. Wallace, Henry, *Uncle Henry's Letters to the Farm Boy*, 80-81.

96. Lord, 138.

97. Wallace, Henry Cantwell. "The Rock Upon Which He Built," February 23, 1917. In *Tributes*, 39.

98. The Wallace Publishing Company. *Tributes*, 41.

99. Ibid., 56.

100. Ibid.

101. Wallace, Henry, *Uncle Henry's Own Story*, vol. 3, 77.

102. Ogilvie, William Edward, *Pioneer Agricultural Journalists* (Chicago: Arthur G. Leonard, 1927), 110.

103. Wallace, Henry, *Uncle Henry's Own Story*, vol. 3, 90.

104. Ingham, Harvey. "Henry Wallace as Editor and Publisher." In *Tributes*, 35.

105. Lord, 163.

106. Ibid., 164.

107. *Des Moines Capital*, "No Bible But a Copy of *Wallaces' Farmer*." In *Tributes*, 17.

108. Quick, 64.

109. Baker, Ray Stannard. "A Leader with No Ulterior Motives," In *Tributes*, 225.

110. Boyle, James. E, "Our Three Henry Wallaces," *The American Mercury*, 34, (1935): 319-27.

111. Thorne, Clifford. "An Unselfish Publicist," Address at the memorial services of the YMCA of Des Moines, 1916. In *Tributes*, 26.

112. Lord, 160.

113. Ibid., 161.

114. Ibid., 160.

115. Ogilvie, 100.

116. Lord, 7.

117. Ibid., 4.

118. Ibid., 7.

119. Ibid., 6.

120. Ibid., 8

121. Ibid., 10.

122. Ibid., 108.

123. Ibid., 11.

124. Ibid.

125. Ibid., 12.

126. Ibid.

127. Sayre, Laura, "Henry A. Wallace," *Newfarm.org,* Rodale Institute, 2005, http://www.newfarm.org/features/0904/wallacecenter/bio.shtml (accessed December 7, 2006).

128. Lord, 217.

129. Ibid., 324.

130. Wallace, Henry, *Uncle Henry's Own Story*, vol. 3, 111-12.

131. Beckman, 48.

132. Thorne, 26.

133. Bailey, Liberty Hyde. "A Tower of Strength to the Country Life Commission." In *Tributes*, 47.

134. Wallace, Henry, *Uncle Henry's Own Story*, vol. 3, 113.

135. Ibid., 114.

136. Ibid.

137. Likely "Just In Case"

138. "Scrap Heap for Boys" originally appeared as the opening chapter in *Letters to the Farm Folk*, and is deployed here as an opening epistle in an editor's-choice grouping of cautionary tales in letter form. The section title "Hardmans, Hardups, and Richmans: Cautionary Tales for the Farm Hobbledehoy" has been added for clarity at the editor's discretion. The term *hobbledehoy*, a favorite of Wallace's, is an archaic term signifying an unintentionally clumsy but well-meaning young man.

139. Two letters from *Letters to the Farm Boy*, "The Brodhead Family" and "Types of Common People" have been fused at this paragraph break for the sake of concision and readability. Both letters have been excerpted and run here at less than half of their original length. Specifically, pages 174 to 186 from "The Brodhead Family" have been excised and 190-202 from "Types of Common People" have been omitted.

140. The final sentence to this preface prepared by the Wallace sons has been omitted here to avoid confusion, as three separate volumes of *Uncle Henry's Own Story of His Life* were eventually published. The excised sentence reads: *The present volume contains all the letters published up to the autumn of 1917.*

141. The original note here read "Uncle Henry's first great-grandchild, a boy, appeared September 18, 1915, some five years after this letter was written." Thus, the address to his great-grandchildren was, at the time, purely anticipatory.

142. Uncle Henry's opening note to his great-grandchildren continues for several additional sentences, which have been excised here in the interest of readability. In the last sentence, Uncle Henry urges his great-grandchildren toward "right living," "faith," and "regard to our treatment of our fellow men."

143. The salutation "My Dear Great-Grandchildren" has been added here and throughout this section in keeping with the salutation of the first letter of the first volume.

144. The last paragraph of this letter, in which Uncle Henry discusses money and currency of the era, is something of a departure and has been excised for the sake of readability.

145. At this juncture the letter reaches a section break before turning to Uncle Henry's pioneering, along with Iowa State Agricultural College's Professor Perry G. Holden, the educational "Seed Corn Trains" of 1904. This latter section has been excised in keeping with the theme of this section: youth.

146. The epigraph "Know ye not that there is a prince and a great man fallen this day in Israel?" originally preceded this first sentence.

147. The following note preceded the appearance of these student essays as they appeared in *Tributes to Henry Wallace*: "The following compositions from pupils of Wallace School are taken from the *Wallace-Whittier Watchword.*"

148. Excerpted from the *Des Moines Capital*, as noted in *Tributes to Henry Wallace*, where the last will ran in full. The opening paragraph of the original *Capital* article, excised here, read, "The will of the late Henry Wallace was filed in district court today. Mr. Wallace departs from the usual custom to talk in a heart-to-heart manner to his children." Of the four numbered directives intended for his children and grandchildren, only items three and four appear in this excerpt.